Environmental, Social, and Governance (ESG) Investment and Reporting

Michaela Bednárová • Kullapa Soratana
Editors

Environmental, Social, and Governance (ESG) Investment and Reporting

Springer

Editors
Michaela Bednárová
Department of Financial Economics and Accounting
University Pablo de Olavide
Seville, Spain

Kullapa Soratana
Faculty of Logistics and Digital Supply Chain
Naresuan University
Muang, Phitsanulok, Thailand

ISBN 978-3-031-84237-5 ISBN 978-3-031-84235-1 (eBook)
https://doi.org/10.1007/978-3-031-84235-1

This Springer imprint is published by the registered company Springer Nature Switzerland AG
The registered company address is: Gewerbestrasse 11, 6330 Cham, Switzerland

Preface

The world is changing rapidly, and the need for businesses and investors to be more transparent about their impact has never been more urgent. Environmental, Social, and Governance (ESG) reporting has become an essential way for companies to share information about how they operate and how they influence the world around them. By using clear frameworks and standards, ESG reporting ensures consistency and comparability, giving investors and stakeholders the tools to evaluate risks and opportunities. Beyond transparency, incorporating ESG factors into investment strategies allows investors to align financial goals with their values, support sustainable practices, and encourage companies that are better prepared for long-term success and positive contributions to society.

This book, *ESG Investment and Reporting*, is a guide designed to help readers understand ESG from multiple perspectives. It begins by introducing the core concepts of ESG, explaining its importance for sustainable development and investment decisions. Ethical investing is also discussed, showing how ESG principles can be integrated into decision-making to align with both personal and professional values. The book explores the practical aspects of ESG, starting with environmental issues like climate risk and the use of tools such as Life Cycle Assessment (LCA) to measure sustainability efforts. Social and governance factors are addressed as well, with a focus on how elements like board diversity, stakeholder relationships, and leadership practices shape business outcomes and investment strategies. Readers will also learn how to apply ESG factors to financial analysis, identify key risks and opportunities, and use data metrics and ratings to make informed decisions.

Finally, the book examines the evolving landscape of sustainable finance and reporting. It covers impact investing, financial instruments that support sustainability, and global ESG reporting frameworks like global reporting initiative (GRI), integrated reporting (IR), sustainability accounting standards board (SASB), international sustainability standards board (ISSB), and european sustainability reporting standards (ESRS). It also highlights challenges like greenwashing and introduces emerging ideas such as environmental, social, governance, digital (ESGD) reporting and Corporate Digital Responsibility, which link sustainability to technology. Whether you are a student, an investor, or a business leader, this book aims to equip you with the knowledge and skills to make ESG a meaningful part of your approach. May this book inspire you to understand ESG as a foundation for advancing a more equitable, conscientious, and sustainable world.

Seville, Spain Michaela Bednárová
Phitsanulok, Thailand Kullapa Soratana

Acknowledgments

We would like to express our sincere gratitude and appreciation to the Erasmus+ KA2 program for their invaluable support and funding toward the implementation of the project titled "Master Studies in Sustainable Development and Management." This project has been an instrumental initiative in promoting international cooperation and enhancing the quality of education in the field of sustainable development. We would like to extend our heartfelt thanks to the European Commission and the Erasmus+ program for their vision and commitment to fostering cross-cultural learning opportunities. The financial assistance provided has played a crucial role in facilitating the development and delivery of this innovative Master's program.

We are particularly grateful to the internal and external reviewers of the MASUDEM project: Melinda Ratkai from Hotelschool, The Hague (The Netherlands), Vatcharapol Sukhotu from the Faculty of Logistics and Digital Supply Chain at Naresuan University (Thailand), and Danai Tanamee from Srinakharinwirot University (Thailand), for their invaluable insights and feedback.

We are also incredibly grateful to Springer Publisher for the opportunity to share this knowledge with a wider audience. Additionally, we would like to thank the anonymous Springer reviewers for their constructive feedback, which helped refine and strengthen the quality of this work. Their belief in the importance of ESG and sustainable finance has enabled this work to reach those who are eager to understand and implement these crucial concepts in their professional and academic pursuits.

To everyone who offered their guidance, feedback, and encouragement, thank you. This book is a reflection of collective effort, and we are truly appreciative of the support we have received along the way.

Lastly, we dedicate this book to all those who are committed to advancing ESG investment and reporting as tools for driving positive change. Your efforts to integrate environmental, social, and governance factors into decision-making not only promote more responsible business practices but also contribute to a more equitable and sustainable global economy. Through your dedication, you are helping to shape a future where financial growth aligns with the well-being of people and the planet, creating lasting value for generations to come.

Contents

About the Authors

Michaela Bednárová Dr. Michaela Bednárová is an Associate Professor at the University Pablo de Olavide in Spain. She earned her Ph.D. in Management and Economics from the University of Huelva. During her doctoral studies, she was a visiting researcher at Rutgers Business School and the University of Economics in Bratislava, as well as a lecturer at the University of Edinburgh.

Her expertise lies in Digital Accounting, Corporate Social Responsibility (CSR), Sustainability, Integrated Reporting, and Corporate Digital Responsibility. In 2021, she was awarded the Emerald Literati Award for her research on blockchain implications in accounting. She is the author of Digital Disruption and Corporate Digital Responsibility and four other books.

Dr. Bednárová holds an editorial role as Associate Editor for the International Journal of Digital Accounting Research. She has participated in various research and Erasmus+ projects, including REPESEA and CHAISE, and currently serves as the local coordinator for the MASUDEM (Master in Sustainable Development and Management) project.

She is also active in professional organizations, serving as a member of the committee for new technologies in accounting at Spanish Accounting and Business Administration Association - Asociación Española de Contabilidad y Administración de Empresas (AECA) (Madrid) and as a member of the Academic Advisory Body of International Association of Trusted Blockchain Applications - International Association for Trusted Blockchain Applications (INATBA) (Brussels).

Kullapa Soratana Dr. Kullapa Soratana, Associate Professor at Naresuan University, Thailand, specialized in life cycle assessment, sustainability, and environmental science. She earned her Ph.D. in Civil and Environmental Engineering, focusing on Sustainability and Green Design, from the University of Pittsburgh, USA, in 2012. With a master's in Environmental and Hazardous Waste Management and a bachelor's in General Science from Chulalongkorn University, Thailand, Dr. Soratana demonstrates a strong commitment to sustainability. Her significant contributions include diverse academia topics, including life cycle assessment and sustainable tourism development. Published in prestigious journals like "Nature Nanotechnology," "Environmental Science & Technology," and "Journal of Cleaner Production," her research attests to her expertise. As an educator, she imparts knowledge through courses such as "Sustainable Logistics and Supply Chain" and "Life Cycle Thinking – A pathway to a sustainable supply chain."

Marisa Laokulrach Assistant Professor Dr. Marisa Laokulrach received her bachelor's degree in Business Administration (majoring in Finance and Banking), and MBA from Assumption University. She obtained Ph.D. in Public Administration (Development Management) from the National Institute of Development Administration.

Currently, Marisa is a Vice Dean of Academic Affairs and a full-time lecturer at the International College of National Institute of Development Administration (ICO NIDA), Thailand. She has been working at ICO NIDA for 6 years with teaching experience in the subjects of Financial Management, and Business and Strategic Management for master's degree program, and Budgeting and Financial Management for Ph.D. program.

She had been a full-time lecturer at Martin de Tours School of Management and Economics, Assumption University, for 8 years before joining ICO NIDA. She is also a guest lecturer for finance-related subjects in different countries for instance Chongqing Technology and Business University (CTBU) in China, Arshad Ayub Graduate Business School, Universiti Teknologi

MARA (UiTM) in Malaysia, and RheinMain University of Applied Sciences in Germany.

Her teaching areas include Financial Investment and Portfolio Management, Financial Management, and Budgeting.

Her research areas include Investment and Portfolio Management, Equity Market, Corporate Financial Management, and Sustainable Finance.

Marta Miranda García Dr. Marta Miranda García holds a Ph.D. in Economics and Business Administration from Rey Juan Carlos University. She received her bachelor's degree in Economics from the Complutense University of Madrid. She is currently an Associate Professor at the University Pablo de Olavide in Spain. Previously, she was an Associate Professor at Rey Juan Carlos University (URJC) in the Department of Financial Management from 2014 to 2021. She also served as an Associate Professor at the Complutense University of Madrid (UCM) in the Department of Financial Management from 1999 to 2012 and as a professor at the Centre for Advanced Studies Felipe II (UCM) from 2000 to 2016. Additionally, she held the position of Sub-Director of the Cooperative School at the Complutense University of Madrid until 2010.

Her expertise focuses on participatory enterprises, particularly cooperatives. Her main research areas include cooperative models, artificial intelligence applied to business and finance, and small business and microfinance. More broadly, her research explores the use of artificial intelligence techniques to address economic and financial challenges.

Nataliia Tkalenko D.Sc., Prof. Nataliia Tkalenko is an accomplished Investment and Project Management professional with a rich academic background. Over the course of 25 years, she has devoted her time to extensive research, academic study, resulting in more than 80 publications.

In addition to her academic accomplishments, Nataliia Tkalenko possesses practical expertise in Project Management. Her professional experience enhances her ability to bridge theoretical knowledge with real-world applications.

Professor Tkalenko is the author of three textbooks, notably "Investment Management," which has gained widespread adoption in Ukrainian universities. Her publications are recognized for their contribution to the field and their relevance in academic settings.

Notably, in 2018, Professor Tkalenko received the prestigious title of Honoured Worker of Science and Technology through the Decree of the President of Ukraine, acknowledging her outstanding achievements and dedication to her field.

Furthermore, Professor Tkalenko actively participates in international research projects, demonstrating her commitment to global academic collaboration.

Nataliia Tkalenko continues to be deeply engaged in her research endeavors, ensuring that her work remains at the forefront of academic exploration and contributing to the global discourse in her field.

Chapter 1
Introduction to ESG Approach

Nataliia Tkalenko

Abstract In the contemporary business landscape, the significance of environmental, social, and governance (ESG) factors has become increasingly pronounced, marking a shift in how companies are evaluated beyond traditional financial metrics. ESG, standing for environmental, social, and governance, is integral to the sustainable development of commercial activities. It encapsulates three interconnected pillars that shape an organization's approach to sustainability and responsible business conduct. The environmental aspect evaluates a company's impact on ecosystems, natural resources, and climate change. The social dimension considers the company's interactions with its employees, customers, communities, and society at large. The governance facet focuses on internal decision-making processes, including board structure, executive remuneration, and shareholder rights.

The increasing emphasis on ESG is driven by its potential to create long-term value and mitigate risks. Firms that integrate ESG principles are often observed to enhance operational efficiency, minimize regulatory risks, and bolster their brand reputation. Furthermore, ESG-centric businesses are more adept at attracting and retaining investors focused on sustainable returns and ethical compliance. In recent years, the business environment has undergone significant changes, with an increasing emphasis on environmental, social, and governance (ESG) factors. ESG has emerged as a framework that goes beyond traditional financial metrics to assess a company's performance and its broader impact. This section explores the fundamental concepts and principles of ESG, shedding light on its importance in the modern global economy and the ethical aspects of ESG. This chapter delves into the core concepts and principles of ESG, highlighting its pivotal role in sustainable development and its implications for investment practices.

N. Tkalenko (✉)
Department of Management, Mendel University in Brno, Brno, Czech Republic
e-mail: nataliia.tkalenko@mendelu.cz

M. Bednárová, K. Soratana (eds.), *Environmental, Social, and Governance (ESG) Investment and Reporting*, https://doi.org/10.1007/978-3-031-84235-1_1

1.1 ESG Concepts and Principles

The abbreviation ESG stands for "environmental, social, and governance." Broadly, this refers to the sustainable development of commercial activities based on the following principles: responsible treatment of the environment, high social responsibility, and high-quality corporate governance. ESG encompasses three interrelated pillars that define an organization's approach to sustainability and responsible business practices. Environmental factors assess a company's impact on ecosystems, natural resources, and climate change. Social factors address the company's relationships with its employees, customers, communities, and the broader society. Governance factors focus on the internal mechanisms shaping decision-making processes, including the structure of the board of directors, executive remuneration, and shareholder rights.

The ESG investment concept is centered around the integration of environmental (E), social (S), and governance (G) factors into investment decisions. The connection between these three components lies in their collective focus on assessing and managing the broader impact and sustainability practices of a company or investment (Fig. 1.1).

Environmental component evaluates how a company's operations impact the natural environment. It includes considerations like carbon emissions, waste management, resource conservation, and the company's response to climate change. Investments that score well on environmental factors are seen as better positioned for long-term sustainability, particularly as the world increasingly focuses on mitigating climate change and transitioning to a greener economy.

Social aspect examines how a company manages relationships with its employees, suppliers, customers, and the communities where it operates. It includes labor

Fig. 1.1 Concept of ESG investment. (Author's elaboration, based on Napoletano and Katzeff (2021))

practices, diversity and inclusion, community engagement, human rights, and customer satisfaction. The social dimension is crucial for understanding a company's social license to operate and how its practices might affect brand reputation and customer loyalty.

Governance refers to a set of rules or practices that dictate how a company is administered or controlled. This includes issues such as board composition and diversity, executive pay, audits, internal controls, and shareholder rights. Strong governance can help prevent scandals, fraud, and other activities that could damage a company's reputation and financial performance.

The connection between E, S, and G is founded on the premise that companies which effectively manage these aspects are more likely to be sustainable and ethical in the long term, thereby potentially offering better risk-adjusted returns to investors. This interconnected approach recognizes the following.

Environmental practices of a company can affect its social relationships and governance structures. For example, a company that neglects environmental considerations may face community backlash or regulatory action, which in turn can impact its governance and overall sustainability.

Social practices are often closely linked with environmental and governance issues. For instance, a company's labor practices can affect employee morale and productivity, which in turn influences overall business performance. Similarly, poor social practices can lead to governance challenges, such as regulatory penalties or legal issues.

Governance structures can influence how well a company manages its environmental and social responsibilities. Effective governance can ensure that a company takes a proactive approach to environmental conservation and social welfare, embedding these priorities into its strategic decision-making processes.

The connection between E, S, and G in ESG investment is that they are interdependent and collectively contribute to a holistic understanding of a company's operational impact, ethical practices, and long-term sustainability. By integrating ESG factors, investors and stakeholders can better assess the risks and opportunities associated with investments, leading to more informed decisions that align with sustainable and ethical business practices. This approach ultimately supports the transition toward a more sustainable global economy.

The growing recognition of the importance of ESG is driven by its potential to create long-term value and reduce risks. Companies integrating ESG principles often experience improvements in operational efficiency, reduced regulatory risks, and strengthened brand reputation. Moreover, ESG-oriented enterprises are better at attracting and retaining investors who are seeking sustainable returns and ethical compliance.

The principles of ESG were first mentioned in the 2004 report of the United Nations Global Compact, prepared with the support of 18 financial institutions from 9 countries around the world (with a total of assets under management of more than six trillion dollars). The authors of the report acknowledged that taking into account ESG factors contributes to the creation of stronger and more sustainable investment markets, as well as to the sustainable development of society as a whole (Who Cares Wins, 2004).

Further development of ESG principles occurred in the work "Principles for Responsible Investment" (PRI), conducted in 2005 under the leadership of the UN Secretary-General and a group of major institutional investors. This initiative was first presented in April 2006 at the New York Stock Exchange.

The number of companies officially recognizing the "Principles for Responsible Investment" has increased from 63 in 2006 to 3826 in 2021, and the total value of assets managed in accordance with ESG principles during this period grew from 6.5 trillion to 121.3 trillion dollars (About the PRI, 2021). The growing interest of investors in ESG factors is driven by the understanding that risks associated with them can impact the activities of issuers in the long term. The Principles for Responsible Investment were developed by an international group of institutional investors in response to the growing number of environmental, social, and governance issues for investment practices. The principles were formulated as shown in Fig. 1.2.

These principles oblige investors to adopt and implement them where it aligns with the fiduciary duties of companies. Over time, these principles may evolve in response to societal demands. They aim to enhance the ability of companies to fulfill their obligations to beneficiaries and better align investment activities with the broader interests of society. Institutional investors bear a special responsibility for acting in the best long-term interests of their beneficiaries (Schanzenbach & Sitkoff, 2020).

In this fiduciary role, investors believe that environmental, social, and governance (ESG) issues can impact the performance of investment portfolios (over time and to varying degrees across different companies, sectors, regions, and asset

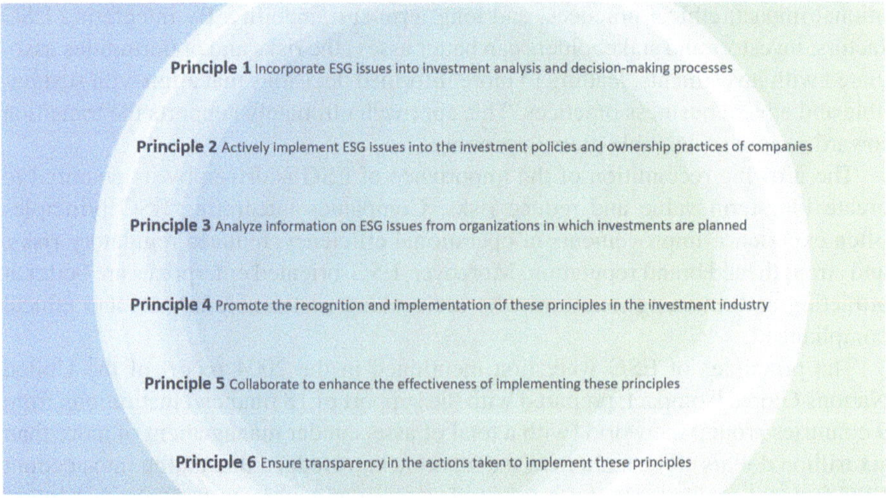

Fig. 1.2 Principles for ESG investment. (Author's elaboration, based on About the PRI (2021); What are the Principles for Responsible Investment? (2021))

classes). The implementation of these principles involves a set of possible actions (Table 1.1).

ESG principles are followed to attract investments (alongside financial indicators of the enterprise, business management standards, environmental pollution risks, and fulfillment of social obligations to employees are also considered); for implementing PR strategies (a good ESG campaign can become a notable news event, while a gross violation of these rules can lead to a serious PR crisis); to increase loyalty of sophisticated customers (an ESG strategy is a component of modern marketing; a responsible approach to ecology, e.g., attracts customers who consider care for the environment important); and for employer branding development (the social component of ESG directly impacts a company's position in the labor market. The better the conditions provided by the employer, the more interesting it is to job seekers).

Implementing ESG principles is not without challenges. Measuring nonfinancial metrics, balancing short-term and long-term goals, and harmonizing ESG standards are ongoing challenges. Nevertheless, the momentum around ESG is undeniable. Under pressure from governments, investors, and consumers, companies are forced to implement sustainable practices.

The concepts and principles of ESG have led to a paradigm shift, transforming companies from mere profit generators into agents of positive change. By taking responsibility for the environment, social justice, and transparent governance, companies can pave the way for a more sustainable and equitable future. The path to excellence in ESG requires dedication, adaptation, and a collective spirit, but the rewards are not limited to financial ones—they include long-term positive impact on society and the planet.

1.2 The Importance of ESG Approach in Sustainable Development and Its Implications for Investing

Environmental, social, and governance (ESG) investing has become a significant aspect of sustainable finance, evolving from socially responsible investment philosophies into a distinct form of responsible investing. This evolution reflects the growing investor interest in incorporating ESG factors into asset allocation and risk decisions, aiming for sustainable long-term financial returns.

The current state of ESG investing is characterized by a wide variety of metrics, methodologies, and approaches. These diverse practices contribute to disparate outcomes and interpretations regarding the performance of high-ESG portfolios.

Several factors have contributed to the increased attention to ESG criteria and investing. These include industry and academic studies suggesting that ESG investing can improve risk management and provide competitive returns, growing societal awareness of environmental and social issues, and the momentum among corporations and financial institutions to adopt longer-term perspectives of risks and returns

Table 1.1 The implementation of principles for ESG investment

Principle 1: Incorporation of ESG issues into investment analysis and decision-making processes	
Possible actions	Considering ESG issues in investment policy statements Supporting the development of tools, metrics, and analyses related to ESG Assessing the capabilities of internal investment managers in incorporating ESG issues Assessing the capabilities of external investment managers in incorporating ESG issues Requesting that investment service providers (such as financial analysts, consultants, brokers, research firms, or rating agencies) integrate ESG factors into their research and analysis Encouraging academic and topic-specific research Promoting ESG education for investment professionals
Principle 2: Active incorporation of ESG issues into company investment policies and equity ownership practices	
Possible actions	Developing and disclosing an active ownership policy aligned with the principles Exercising voting rights or monitoring adherence to a voting policy (if delegated) Developing the ability to engage (directly or through third parties) Participating in the development of policy, regulation, and standard setting (e.g., in support and protection of shareholder rights) Filing shareholder resolutions consistent with long-term considerations in ESG Engaging with companies on ESG issues Participating in collaborative engagement initiatives Requesting investment managers to conduct and report on ESG engagement
Principle 3: Analysis of ESG-related information from organizations targeted for investment	
Possible actions	Requesting standardized reporting on ESG issues (using tools such as the Global Reporting Initiative) Requesting the integration of ESG issues into annual financial reports Requesting information from companies regarding their adoption/compliance with relevant standards, codes of conduct, or international initiatives (e.g., the United Nations Global Compact) Supporting shareholder initiatives and resolutions that promote the disclosure of information on ESG issues
Principle 4: Promote the recognition and implementation of these principles within the investment industry	
Possible actions	Incorporating principles-related requirements in requests for proposals (RFPs) Aligning investment mandates, monitoring procedures, performance metrics, and incentive systems accordingly (e.g., ensuring investment management processes reflect long-term time horizons when appropriate) Communicating ESG expectations to investment service providers Reassessing relationships with service providers that do not meet ESG expectations Supporting the development of tools for comparing ESG integration Supporting regulatory or policy initiatives that facilitate the implementation of the principles

(continued)

Table 1.1 (continued)

Principle 5: Collaborate to enhance the effectiveness of implementing these principles	
Possible actions	Supporting/participating in networks and information platforms for sharing tools, pooling resources, and using investor reporting as a learning source Collectively addressing relevant issues Developing or supporting appropriate joint initiatives
Principle 6: Transparency in the implementation of these principles	
Possible actions	Disclosing how ESG issues are integrated into investment practices Disclosing active ownership actions (voting, engagement, and/or policy dialogue) Disclosing requirements for service providers in relation to the principles Reporting to beneficiaries about ESG issues and the principles Reporting on progress and/or achievements related to the principles using an "apply or explain" approach Aiming to assess the impact of the principles Make use of reporting to raise awareness among a broader group of stakeholder

(Eccles et al., 2012). Moreover, the public sector, including central banks, has shown interest in transitioning financial systems toward greener, low-carbon economies, integrating ESG assessments into their responsibilities.

In response to this growing demand, the finance industry has expanded its offerings related to ESG ratings, indices, and funds. The market has witnessed a proliferation of ESG-related products and services, catering to investors aiming to position themselves for a transition to a low-carbon economy. However, this growth has also invited scrutiny, highlighting the need for ESG investing practices to evolve to meet user expectations and sustain trust.

Sustainable finance, in general, is driven by investors' desires to have an environmental and social impact alongside economic performance. ESG investing specifically aims to combine better risk management with improved portfolio returns. It incorporates long-term financial risks and opportunities, including environmental, social, and governance challenges, into investment decision-making processes.

The integration of ESG factors into investment analyses and decision-making processes is a developing area in finance. The CFA Institute and Principles for Responsible Investment (PRI) have provided guidance and case studies for ESG integration in both equities and fixed income portfolios. This endeavor involved surveying financial professionals, conducting workshops, interviewing practitioners, and analyzing ESG company disclosure scores.

ESG integration is defined as the explicit and systematic inclusion of ESG factors in investment analysis and investment decisions. It is a comprehensive approach where both ESG and traditional financial factors are identified and assessed to form investment decisions. The process typically involves three components: research, security and portfolio analysis, and the investment decision itself.

In equity analysis, ESG integration has historically been associated with fundamental strategies. Practitioners use qualitative and quantitative analyzes to inform investment decisions, including adjusting future revenue growth rates, operating costs, capital expenditures, discount rates, and terminal values based on ESG factors. This approach is evolving to include systematic strategies like quant and smart beta strategies, where ESG factors are integrated into valuation models and investment decisions.

Fixed-income investing, compared to equity investing, is still in the early stages of integrating ESG factors. Traditionally, corporate bond practitioners adapted the materiality/sustainability frameworks and ESG techniques used by equity practitioners. However, recently, ESG integration techniques in fixed-income have become more sophisticated, with practitioners adapting their processes and analyses to incorporate ESG factors more fully.

ESG integration is an evolving field in investment management, with increasing emphasis on incorporating ESG factors into both equity and fixed-income investing. The approach involves a combination of qualitative and quantitative analyzes to assess the impact of ESG factors on investment performance and risk. As the field develops, the methods and practices of ESG integration continue to become more refined and integrated into the broader investment process.

The popularity of ESG investments is growing every year. Experts believe this is partly due to the interests of millennials (born in the 1980s and 1990s), who have become a financially capable audience. The values of this generation differ from the previous one: for them, business and investments are not just about income, but also about caring for the environment and society.

Amid the pandemic, there has also been an increased demand for corporate social responsibility. According to an Edelman study, 71% of consumers are ready to abandon a brand if it prioritizes profit over caring for people (2020 Edelman Trust Barometer Brands and the Coronavirus, 2020).

The heightened demand for ESG forces companies to adhere to sustainable development principles. Investors are less supportive of companies with low ESG ratings. In 2020, EY conducted a survey among institutional investors—insurance and investment companies, pension, and charitable funds. As a result, 98% of respondents stated that they strictly monitor a company's ESG rating. Here are the main reasons: a positive correlation between responsible investing and securities performance has been confirmed.

Focusing on ESG ratings, investors can avoid companies whose activities are associated with environmental risks and major financial losses, for example, the oil spill due to the explosion of Transocean's platform in 2010. Banks consider ESG ratings when issuing loans. There are many examples of successful corporations worldwide that adhere to ESG principles. Corporate Knights annually publishes a ranking of the 100 most sustainable global corporations in the world, based on publicly disclosed data. According to Corporate Knights, in 2021, the top five included: French engineering company Schneider Electric, Danish multinational energy company Ørsted A/S, National Bank of Brazil Banco do Brasil SA, Finnish oil and gas company Neste, and international professional services company in design, architecture, and consulting Stantec Inc.

Investors also prefer different ESG investment strategies for themselves:

1. Thematic investments. Some choose to invest in green companies. For example, some finance the construction of solar panels, while others invest in wastewater treatment. This pays off: the value of such European firms grew by 448% from 2007 to 2015.
2. Positive screening. Investors support companies that adhere to ESG principles better than others in their economic sector.
3. Negative screening. Investors refuse to invest in specific companies or sectors that do not support ESG principles, such as producers of alcohol, tobacco, or weapons.
4. Active engagement. Some investors actively influence the companies they invest in: they demand the adoption of resolutions and conduct votes on environmental and social issues, even changing management for better governance. ESG companies appreciate in value because they are supported by major investors.

Institutional investors—insurance companies, funds, and banks—are the most active in promoting the ESG agenda: 91% of them are developing ESG investment strategies. Such organizations manage tens of billions of dollars as they handle money from thousands of private depositors. Thanks to institutional investors, the volume of capital in responsible investments has exceeded 30 trillion dollars. Some investors create ESG organizations like PRI, GSIA, GIIN, or IIGCC. The most active association is Climate Action 100+: it consists of more than 500 investors who manage $55 trillion. They can push a company's shares up simply by the fact of their investment.

The energy company NextEra grew by 443% over 10 years, largely because in 2011, it announced it would primarily generate electricity from renewable sources: wind and solar. Governments are developing laws and guidelines that compel companies to disclose information about ESG indicators. This makes it easier for investors to make decisions. The European Union has adopted a classification of climate activities, a directive on sustainable development reporting rules, and a requirement to consider ESG factors in financial advice. The UK is preparing requirements for the disclosure of climate-related financial information and recommendations on publishing nonfinancial ESG information. The USA is discussing guidelines to disclose information about environmental impact, gender equality, and political lobbying expenses. They are also preparing requirements to consider ESG factors in assessing risk and return on investments. China recommends investors to take into account ESG factors and mandates companies to publish them in their reports. Governments also support ESG companies by offering tax exemptions or reductions. They also invest in scientific research and development or subsidize products.

Modern investors recognize the advantages of ESG investing (i.e., investing considering environmental and social factors, as well as corporate governance factors) and expect corresponding reporting on these parameters. In 2020, the demand for sustainable investments outpaced traditional ones. According to a study by the Luxembourg Fund Industry Association (ALFI), more than half of the new subscriptions were for ESG funds. Let's consider five ESG investing trends that, according to our forecasts, will manifest in the global economy after the coronavirus pandemic:

- *Special attention to social factors (Trend 1)* The COVID-19 pandemic has uncovered a layer of systemic social issues, bringing them to the forefront of public discourse. Previously, these issues were lower in priority compared to the environmental agenda, as they are harder to define and assess. When it comes to improving financial sustainability, social risks should not be underestimated—their impact can be significant. For example, complying with occupational health and safety requirements requires certain initial investments from entrepreneurs, but ultimately reduces the risk of costly lawsuits. Therefore, it is particularly important for ESG investors to consider diversity and inclusivity, as well as current social issues.
- *Improved collection of ESG data in response to new reporting requirements (Trend 2)* The quality and volume of ESG data will continue to grow as a result of stricter corporate reporting requirements. The current heterogeneity of reporting standards complicates the process of collecting and comparing data on ESG progress for investors and makes it difficult to respond to emerging risks and opportunities in this area. Unified requirements will help simplify this process. Regulatory frameworks around the world have faced the challenge of a lack of clear definitions and standardized data. In Europe, the Sustainable Finance Disclosure Regulation (SFDR) imposes obligations to disclose ESG data, and the EU's "Green" taxonomy presents a list of environmentally sustainable activities. SFDR is designed to increase the transparency of sustainable activities, prevent so-called greenwashing, and ensure informed investment decisions.
- *CO_2 capture and carbon offsets to become mainstream (Trend 3)* Compensating for CO_2 or other greenhouse gas emissions (thanks to carbon offsets) will become mainstream, as these measures offer a short-term solution for emission reduction. Of course, there are traditional methods of capture (planting trees or restoring depleted lands), but now preference is given to more advanced solutions, such as carbon capture and storage technologies, which allow removing CO_2 from the atmosphere.
- *Assessment of the impact on biodiversity (Trend 4)* Biodiversity is rapidly declining worldwide (i.e., the diversity of living organisms on the planet). This can have as serious an impact on life and the economy as climate change. As the COVID-19 pandemic has shown, the trade in wild animals and plants, along with deforestation for agricultural needs or development, can lead to the emergence of new zoonotic diseases. This, in turn, carries catastrophic consequences for the economy. Disclosure of the impact on biodiversity is still a rare practice, but in some countries, investment funds are already required to report on the impact of their corporate activities on biodiversity. In France, for example, financial institutions are required to disclose their strategies for reducing the impact on biodiversity, including specific target indicators and tools to align local strategies with international goals in the field of biodiversity conservation.

1.3 Ethics in Investments: ESG Integration in Investment Decision-Making

More and more private capital owners are contemplating not only the economic but also the ethical aspects of their investments. In this context, socially responsible approaches to investing are becoming relevant:

- "Making Money, Do Not Harm" Approach: This approach involves selecting investment targets that exclude companies whose activities harm the environment and society or violate ethical norms.
- "Do Good While Making Money" Approach: Based on the principle that investments should not only generate income for the capital owner but also benefit society and the environment. Socially responsible investment funds in their modern form emerged in the 1960s, driven by escalating issues such as environmental protection, civil rights, and nuclear energy concerns.

In global practice, there are various definitions of ethical investing or socially responsible investing (SRI), also known as social investing, socially aware investing, ethical investing, mission-based investing, and natural investing. This is a process of making investment decisions that consider social and environmental consequences within the framework of traditional financial analysis. Ethical investments encompass the protection of human and human rights, support for fairness and justice, legality, and enhanced accountability.

Incorporating environmental, social, and governance considerations into investment strategies is becoming critical for investors aiming to match their portfolios with their ethical standards and foster responsible business conduct. ESG elements offer crucial insights into the enduring viability of a company's operations and its capability for delivering appealing financial outcomes. This holistic approach incorporates three primary components: research, security and portfolio analysis, and investment decision-making.

ESG integration in investment decision-making is a comprehensive process that encompasses both qualitative and quantitative analyses. It requires a deep understanding of how ESG factors impact both corporate and investment performance and involves ongoing adjustments to investment strategies based on these factors. This approach reflects a shift in the investment world toward more sustainable and responsible investment practices, considering the broader impacts of business activities on society and the environment.

ESG integration often involves both qualitative and quantitative approaches. Qualitative analysis may include assessing a company or country's ESG performance alongside its financial valuation to inform investment decisions. For instance, a poor ESG performance might lead to a "sell" or "don't invest" decision, or it might prompt a deeper analysis if the financial valuation is strong. Quantitative analysis involves adjusting internal credit assessments, financial forecasts, valuation-model variables, and portfolio weightings based on ESG analysis and scores. Such adjustments can be temporary or permanent, made through sensitivity analysis or scenario planning, and can influence portfolio construction decisions.

The integration of ESG factors differs between equity and fixed-income investing. Equity practitioners typically integrate ESG factors into their valuation models and investment decisions, using internal and third-party research to create ESG scores for companies. These scores are considered alongside other factors like value, size, momentum, growth, and volatility. In contrast, fixed-income practitioners have historically been slower to integrate ESG factors, but recent developments have seen more sophisticated integration techniques being applied, with considerations for the varied nature of bonds.

In fixed-income analysis, practitioners initially adapted ESG frameworks and techniques from equity analysis. However, the analysis of ESG risks and opportunities in fixed-income investing requires a nuanced approach due to the diverse nature of bonds. Factors like issuer types, credit quality, duration, payment schedules, and liquidity risks require a multilayered analysis of credit risk, interest rate risk, yield curve risk, and liquidity risk. Material ESG issues are considered consistent whether the investor is a shareholder or a bondholder, with ESG scores and research being integrated into traditional credit analysis to influence investment decisions.

ESG considerations are instrumental in identifying risks and opportunities that might be overlooked by conventional financial analyses. For instance, companies with deficient labor practices or environmental compliance issues might encounter substantial legal and reputational challenges, potentially affecting their financial health. On the other hand, firms dedicated to sustainable practices and robust governance may gain a competitive edge and enjoy long-term success.

Investors have several methods to integrate ESG considerations. One is utilizing ESG ratings and data from specialized research agencies, which assess and rate companies based on their ESG performance. These ratings allow investors to evaluate and benchmark potential investments. Alternatively, investors might conduct their own ESG assessments using public data such as sustainability reports, regulatory disclosures, and media coverage. Direct engagement with companies is another way to gain deeper ESG insights. Moreover, investors might adopt ESG-focused investment strategies, like sustainable or impact investing, prioritizing companies that lead in sustainability and ESG practices or address significant global challenges.

Integrating ESG elements offers several benefits:

- Enhanced risk management. Identifying potential risks and opportunities beyond the scope of traditional financial analysis leads to more robust risk management.
- Potential for better returns. Companies committed to ethical practices and sustainability may outperform others in the long term.
- Value alignment. ESG integration helps investors align their portfolios with their personal values and support ethical business practices.
- Reduced reputational risk. By focusing on ESG factors, investors can avoid associations with companies engaged in controversial or unethical practices.

Integrating ESG factors also presents challenges, including a lack of uniform ESG metrics and reporting standards. This disparity hinders the ability to compare companies' ESG performances across sectors and geographies and poses challenges in transparent reporting. Another issue is the possible conflict between ESG performance and immediate financial returns. Investments in high-ESG-performing companies might limit opportunities and yield lower short-term returns, though this could be balanced by better long-term financial outcomes and reduced risks. Assessing the direct impact of ESG integration on financial performance is complex. While studies indicate that strong ESG performers often outperform in the long term, quantifying the specific financial impact of ESG factors can be challenging.

1.4 Business Case: BASF's Strategic ESG Transformation

BASF, originally known as Badische Anilin- und Soda-Fabrik, was founded on April 6, 1865, in Mannheim, Germany, by Friedrich Engelhorn. The company started its operations by producing dyes, utilizing the industrial synthesis of aniline from coal tar, a by-product of gasworks. Throughout its history, BASF has undergone numerous transformations, driven by innovations, strategic partnerships, and a commitment to addressing global challenges, securing its position as a leader in the chemical industry.

Today, BASF, one of the world's largest chemical companies, employs a comprehensive and systematic approach to environmental, social, and governance (ESG) issues as part of its corporate strategy. The company's ESG strategy is deeply integrated into its operations, business, and corporate culture. Here are some key highlights of BASF's ESG strategy based on information from their website and recent reports.

BASF's corporate purpose, "We create chemistry for a sustainable future," is grounded in sustainability (www.basf.com). The company aims to contribute to a world that offers a viable future with an improved quality of life for everyone. This is achieved by integrating the three pillars of sustainability—economy, environment, and society—into its corporate strategy, goals, and operations.

BASF analyzes its products' contributions to sustainability, paying special attention to the needs of its customers' value chains. This is part of its Sustainable Portfolio Management program, where BASF assesses its products' applications in relevant markets and customer industries. Products that raise significant sustainability concerns are classified as "challenged," and action plans are developed for these products, including research projects, recipe modifications, or replacements.

The company aims to reduce carbon dioxide emissions and has set ambitious targets to achieve net-zero CO_2 emissions by 2050. This involves using renewable energy sources, developing new low-carbon production processes, and continuously increasing the efficiency of energy and resource use. BASF has established the Net Zero Accelerator unit to implement and accelerate projects focused on low-carbon production technologies, the circular economy, and renewable energy sources.

BASF aligns its actions with the circular economy principle, aiming to close cycles and use products and resources throughout the entire value chain. This approach helps customers achieve their sustainability goals and reduces the material and carbon footprint of BASF's products.

The company ensures responsible procurement of raw materials and maintains high standards for safe and efficient production. This includes managing human rights, environmental protection, and anti-corruption in its supply chains, as well as prioritizing health and safety, emergency response, and product stewardship in its operations.

As a co-founder of the UN Global Compact, BASF contributes to the implementation of the UN's 2030 Agenda. The company's products, solutions, and technologies help achieve several UN goals, including Zero Hunger, Gender Equality, Clean Water and Sanitation, Affordable and Clean Energy, Decent Work and Economic Growth, Responsible Consumption and Production, and Climate Action.

BASF's commitment to ESG principles has significantly influenced its economic ranking and overall performance in various ways. By integrating ESG criteria into its core business strategy, BASF has not only enhanced its sustainability credentials but also strengthened its market position, financial stability, and investor appeal. High ratings from organizations like FTSE4Good, ISS ESG, MSCI ESG, Morningstar Sustainalytics, and Vigeo Eiris reflect the company's strong ESG performance. BASF's ESG strategy, particularly its focus on environmental protection and resource efficiency, has led to more sustainable operations, operational efficiencies, and cost savings. Active risk management and regulatory compliance have mitigated financial losses and safeguarded the company's economic stability and ranking. The company's efforts in sustainability and social responsibility have resonated with customers and the public, leading to enhanced brand loyalty and reputation.

1. How has BASF's commitment to ESG principles influenced its economic ranking and investor appeal?
2. What strategic measures has BASF implemented to align its operations with the principles of the circular economy?
3. In what ways does BASF's ESG strategy address the UN's Sustainable Development Goals?

1.5 Summary

This chapter presents a comprehensive examination of the integration of environmental, social, and governance (ESG) factors into investment strategies, highlighting the varied methodologies and their impact on investment decisions. It elaborates on the different approaches that investors adopt to incorporate ESG considerations, including using specialized ESG ratings and conducting independent evaluations based on public data and direct engagements with companies. The discussion emphasizes the multifaceted advantages of integrating ESG elements, such as improved risk management, the potential for enhanced long-term returns, alignment of investments with personal values, and mitigation of reputational risks linked to unethical corporate behaviors.

The challenges associated with ESG integration are also critically examined. These include the absence of standardized ESG metrics and reporting standards, which complicates the assessment and comparison of ESG performances across various sectors and geographical areas. The chapter addresses the delicate balance between achieving immediate financial returns and maintaining strong ESG performance, acknowledging the possible short-term compromises for long-term benefits and risk mitigation. Additionally, the complexity of quantifying the direct financial impact of ESG factors is discussed, underscoring the need for a sophisticated understanding of both ESG and traditional financial metrics.

Through its thorough analysis, the chapter underscores the increasing significance of ESG considerations in modern investment practices, illustrating how they are reshaping the landscape toward sustainable and ethically aligned investment choices.

1.6 Discussion Questions

1. How do ESG considerations influence the identification of potential risks and opportunities in investment analysis?
2. Discuss the methods investors can use to integrate ESG considerations into their investment strategies. What are the pros and cons of each method?
3. What are the primary benefits of integrating ESG factors into investment decision-making? How do these benefits align with investor goals and values?
4. What challenges do investors face in integrating ESG factors into their investment strategies, and how can these challenges be mitigated?
5. How does the lack of standardized ESG metrics and reporting standards impact the ability to compare ESG performances across different sectors and regions?
6. Explore the potential conflict between ESG performance and immediate financial returns. How can investors balance these aspects in their investment decisions?
7. Discuss the complexity of assessing the direct impact of ESG integration on financial performance. What approaches can be used to evaluate this impact?

References

2020 Edelman Trust Barometer Brands and the Coronavirus. (2020). *Edelma*. Retrieved from https://www.edelman.com/sites/g/files/aatuss191/files/2020-03/2020%20Edelman%20Trust%20Barometer%20Brands%20and%20the%20Coronavirus.pdf

About the PRI. (2021). *Principles for responsible investment*. Retrieved from https://www.unpri.org/pri/about-the-pri

Eccles, R. G., Ioannou, I., & Serafeim, G. (2012). *The impact of corporate sustainability on organizational processes and performance* (NBER Working Paper No. 17950). Retrieved from https://www.nber.org/papers/w17950

Napoletano, E., & Katzeff, P. (2021, March 1). Environmental, social and governance: What is ESG investing? *Forbes*. Retrieved from https://www.forbes.com/advisor/investing/esg-investing/

Schanzenbach, M. M., & Sitkoff, R. H. (2020, October 1) *ESG Investing: Theory, Evidence, and Fiduciary Principles*. Journal of Financial Planning, Available at SSRN: https://ssrn.com/abstract=3684979

What are the Principles for Responsible Investment? (2021). *Principles for responsible investment*. Retrieved from https://www.unpri.org/pri/what-are-the-principles-for-responsible-investment

Who Cares Wins. Connecting Financial Markets to a Changing World. (2004). *The global compact*. Retrieved from https://www.unepfi.org/fileadmin/events/2004/stocks/who_cares_wins_global_compact_2004.pdf

Chapter 2
Environmental Factors in ESG Investing

Kullapa Soratana

Abstract This chapter explores the critical role of environmental factors within the context of environmental, social, and governance (ESG) investing, with a particular focus on climate risk management. As climate-related risks increasingly affect financial markets, understanding and integrating environmental considerations into investment strategies has become essential. In this context, frameworks such as the Task Force on Climate-Related Financial Disclosures (TCFD) are explored for their role in promoting transparency and standardized reporting, which enable investors to assess the potential impacts of climate change on both businesses and portfolios. A key focus is placed on tools like scenario analysis, which are crucial for evaluating the long-term resilience of investments under various climate scenarios. This chapter discusses on the integration of environmental risks into existing enterprise risk management frameworks. It outlines the process of identifying, assessing, and mitigating climate risks, offering a detailed view of how these processes can be embedded into organizational strategies and corporate culture. The chapter also highlights investors' strategies for managing environmental risks, including diversification, active engagement with portfolio companies, green investments, and the use of ESG criteria to identify sustainable opportunities. The growing demand for environmental metrics, such as carbon pricing and greenhouse gas reduction targets, is addressed as investors increasingly seek to align their portfolios with sustainability objectives. Ultimately, the chapter underscores the importance of environmental factors in ESG investing, illustrating how proactive climate risk management can contribute to more resilient and sustainable financial outcomes.

K. Soratana (✉)
Faculty of Logistics and Digital Supply Chain, Naresuan University, Muang, Phitsanulok, Thailand

M. Bednárová, K. Soratana (eds.), *Environmental, Social, and Governance (ESG) Investment and Reporting*, https://doi.org/10.1007/978-3-031-84235-1_2

2.1 Environmental Sustainability in the Investment Context

2.1.1 Sustainability and Sustainable Development

Climate change serves as a significant driving force compelling humanity to actively pursue sustainability. It encompasses long-term alterations in global weather patterns and temperatures, primarily caused by human activities such as fossil fuel combustion, deforestation, and industrial processes. These activities release greenhouse gases into the atmosphere, trapping heat and contributing to global warming. Such climatic shifts have led to a series of severe and catastrophic events, such as El Niño and La Niña, characterized by abnormal ocean temperatures—either warmer or cooler than the average sea surface temperature—resulting in more active hurricane seasons. In June 2023, record-high global temperatures were observed, a phenomenon referred to as "global boiling." A clear and evident outcome of these environmental changes was an unprecedented drought affecting rivers near the Panama Canal, a vital global trade route. This resulted in increased logistics costs, including surcharges and reduced ship weight limits. Achieving sustainability is now an urgent priority to mitigate the consequences of these climatic shifts.

To achieve sustainability, it is essential to maintain equilibrium among the three pillars: the environment (planet), society (people), and the economy (profits), as illustrated in Fig. 2.1. *Sustainability* is defined as "sustainable development that meets the needs of the present without compromising the ability of future generations to meet their own needs" (Brundtland, 1987). Failure to recognize and uphold this equilibrium can lead to significant issues, including socioeconomic inequality, a substantial contributor to environmental degradation. This degradation includes critical challenges such as climate change, natural resource depletion, biodiversity loss, and waste and pollution generation. Society must raise awareness of its

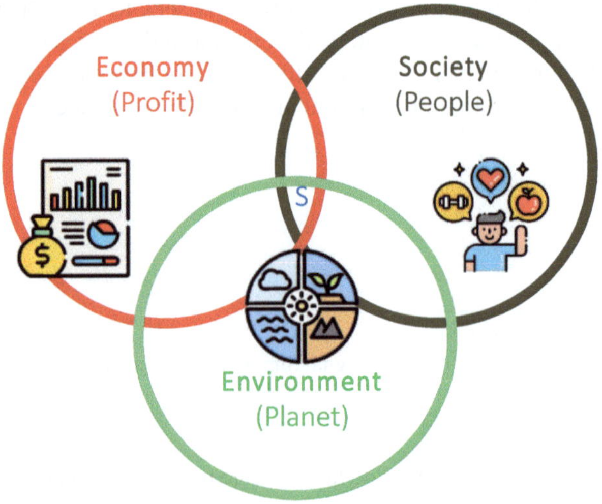

Fig. 2.1 The three pillars of sustainability or the "triple bottom line". (Author's elaboration, based on Elkington (1997))

impacts, create a sense of urgency, and adjust lifestyle to contribute to a more sustainable environment.

In this chapter, we focus on the environmental aspect. Environmental sustainability (planet) is often defined as "bearable," emphasizing the need for human activities to operate within the planet's capacity to avoid harm to ecosystems. The "planetary boundaries" concept highlights nine critical limits for Earth's stability. Staying within these boundaries prevents irreversible damage and ensures sustainable development, while crossing these limits could lead to major environmental shifts that threaten humanity and future generations.

The 2023 update to the planetary boundaries is illustrated in Fig. 2.2. Notably, six out of the nine boundaries have already been crossed, indicating that human activities are pushing Earth's system beyond safe limits. This escalation increases the risk of widespread environmental disruptions, undermining the planet's ability to support life sustainably. Implementing measures to reduce carbon emissions through policies and air pollution limits is crucial to ensuring manageable environmental impacts for present and future generations. Sustainable practices in areas such as agriculture, supply chain management, and other initiatives play vital roles in reversing these trends and achieving environmental sustainability.

The environment holds a vital position in the business and sustainability domains, defined as the Earth and its natural resources, which must be protected and efficiently utilized. According to the System of Environmental-Economic Accounting (SEEA), the environment comprises stocks such as minerals, energy, timber, water, aquatic life, and other biological resources. These stocks give rise to flows of

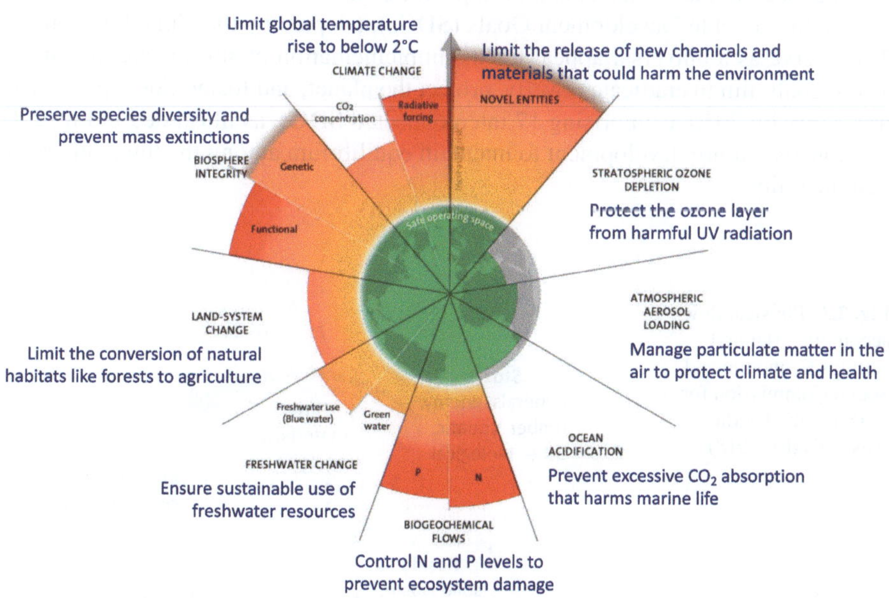

Fig. 2.2 The 2023 update to the planetary boundaries. (Modified from Azote for Stockholm Resilience Centre, based on analysis in Richardson et al. (2023). CC BY-NC-ND 3.0)

materials, energy, and water, which, in turn, support the economy through society's production and consumption processes. As depicted in Fig. 2.3, this economic activity leads to the generation of solid waste, air emissions, and effluents, collectively referred to as residuals, which ultimately return to the environment (United Nations Economic and Social Commission for Asia and the Pacific (UNESCAP), 2018). To ensure the sustainability of our economy for the benefit of society, consumption practices must align with replenishment rates to maintain resource flows and minimize the production of residuals.

In the context of investments, the environment represents more than just a resource. Investors and businesses are increasingly recognizing the value of nature as *natural capital* (International Integrated Reporting Council (IIRC), 2013). This term encompasses both renewable and non-renewable natural assets, including flora, fauna, the atmosphere, water bodies, soils, and minerals, all of which significantly contribute to the well-being of humanity.

In 2015, the Paris Agreement established the framework for worldwide efforts. Almost 200 countries adopted a set of guidelines to steer individual and collective actions aimed at restricting further global warming, which has significant ramifications for both the public and private sectors. These actions are motivated in part by the potential risks to human well-being, infrastructure, natural resources, energy stability, and global stability. However, widely accepted climate scenarios also highlight a crucial economic necessity that must not be overlooked. Nevertheless, achieving sustainability is not solely about "saving the planet"; it involves pursuing strategic, resilient, and sustainable economic growth and development on a global scale, as seen in the Sustainable Development Goals.

The Sustainable Development Goals (SDGs), adopted by the United Nations in 2015, serve as a universal appeal for the implementation of sustainable measures. These goals aim to eradicate poverty, protect the planet, and foster global peace and prosperity by 2030. Comprising 17 interconnected SDGs, as illustrated in Fig. 2.4, these goals require development to maintain equilibrium among the three pillars of sustainability.

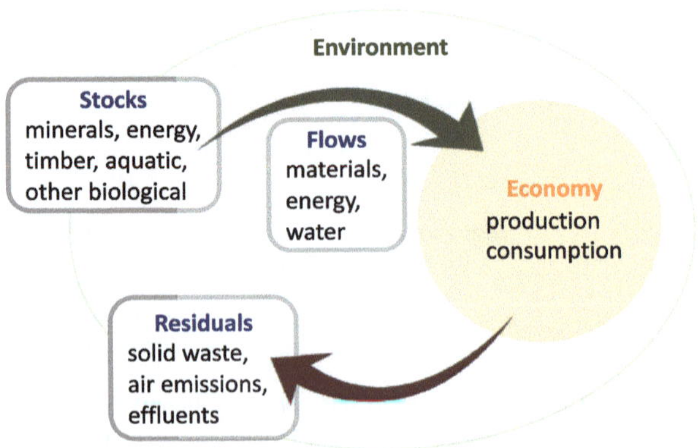

Fig. 2.3 Physical flow accounting. (United Nations Economic and Social Commission for Asia and the Pacific (UNESCAP), 2018)

Fig. 2.4 Sustainable Development Goals. (United Nations, n.d.)

The seven Sustainable Development Goals (SDGs) centered on environmental concerns constitute a comprehensive framework for addressing global sustainability challenges. These goals include ensuring access to clean water and sanitation (Goal 6), promoting affordable and clean energy (Goal 7), fostering sustainable cities and communities (Goal 11), advocating responsible consumption and production (Goal 12), combating climate change (Goal 13), conserving life below water (Goal 14), and protecting and restoring life on land (Goal 15). Collectively, these objectives underscore the pressing need to safeguard the environment, endorse responsible resource management, mitigate the impacts of climate change, and preserve biodiversity. Recognizing that a thriving planet is integral to broader development goals, industries can strategically choose relevant SDGs aligned with their activities. For instance, a company heavily reliant on electricity can contribute to achieving SDG 7 by increasing its utilization of renewable energy sources to ensure accessible and clean energy for society.

2.1.2 *ESG and Sustainability*

ESG (environmental, social, and governance) and sustainability are distinct concepts, yet they share a connection. *Sustainability* encompasses a broader understanding of a company's influence on the environment and society, including issues such as climate change, inequality, and environmental preservation. It involves efforts to both mitigate and remediate any impacts on these issues, making it akin to an "inside-out impact" flow, indicating the impacts an organization has on the planet and society. In contrast, *ESG* is a response from the corporate and financial sectors to societal demands for sustainable development, resembling an "outside-in impact" flow that signifies the impacts of the planet and society on the company. Most companies perform a *materiality assessment* process to identify and describe significant ESG issues relevant to business and stakeholder interests (WBCSD, 2017). The process typically includes a combination of peer benchmarking, megatrends analysis, and engagement with both internal and external stakeholders. The results are often presented in the *materiality matrix*, which visually represents the key ESG issues most relevant to the company. More details on materiality assessment can be found in Chap. 4.

When discussing ESG and sustainability, the concept of *double materiality* is often invoked. Double materiality refers to two perspectives: financial materiality, how sustainability issues impact a company's financial performance, and environmental and social (nonfinancial) materiality, how the company's operations affect the environment and society. It emphasizes the importance of issues and recognizes that what is financially significant to the organization may also hold societal significance, and vice versa, as illustrated in Fig. 2.5.

To advance sustainability goals and effectively engage stakeholders, the corporate sector is increasingly placing a strong emphasis on transparently disclosing information about its sustainability initiatives. The World Business Council for Sustainable Development (WBCSD), founded in 1995, plays an important role in bringing together leading businesses and multinational corporations from various industries to promote sustainability, ESG initiatives, and responsible business practices. Member companies of WBCSD are committed to driving sustainability initiatives, addressing critical global challenges such as climate change, biodiversity loss, and social inequality while advocating for policies and practices that strive to create a more sustainable and equitable world by 2050. This commitment by corporations

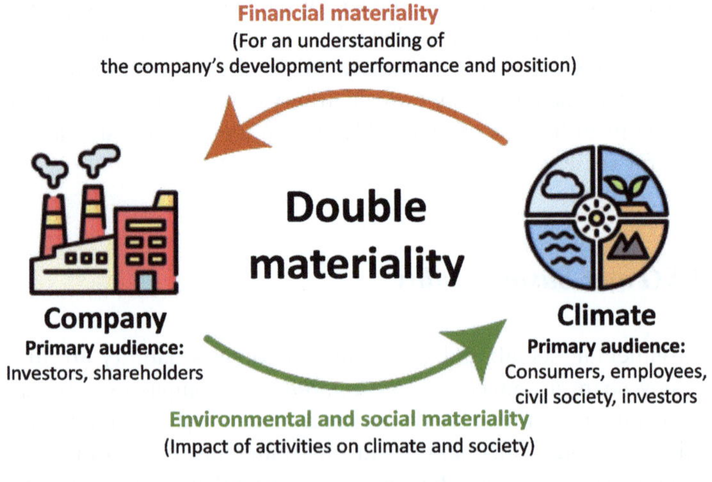

Fig. 2.5 Double materiality indicating the relationship between ESG and sustainability. (Deloitte, 2022)

is demonstrated through sustainability reports, which are designed to showcase how forward-thinking companies utilize the reporting process to drive positive change within their organizations while effectively meeting the needs of their stakeholders.

Sustainability reports, also known as corporate sustainability reports or disclosures, are vital tools for conveying the positive and negative impacts of business operations on society and the environment. These reports provide a comprehensive overview of an organization's sustainability practices and performance, covering areas such as environmental impact, social responsibility, corporate governance, and economic performance. Sustainability reporting encompasses the disclosure of ESG objectives. Through sustainability reporting, enterprises can enjoy a range of advantages, including more effective risk management, streamlined cost structures, an enhanced reputation, and more. These benefits result from identifying social and environmental risks, formulating comprehensive risk management strategies, setting ambitious new goals, and integrating sustainable business practices. This holistic approach empowers businesses to navigate evolving challenges and opportunities with greater agility, thereby promoting a more resilient and responsible path forward.

In the corporate world, another invaluable reporting tool frequently utilized to convey a company's ESG performance and initiatives is *the ESG report*. This specialized report can act as a component within the broader sustainability report, enabling companies to disclose nonfinancial data pertaining to the ESG aspects of their operations. Nevertheless, the Task Force on Climate-related Financial Disclosure (TCFD) recommendations mandate that companies should disclose the potential financial consequences of climate-related risks and opportunities in their primary financial reports. More details on ESG reporting can be found in Chap. 7.

While sustainability reports and ESG reports are often used interchangeably, there are key distinctions between the two. Sustainability reports typically focus on a company's broader environmental and social impacts, while ESG reports provide a more structured framework to assess how an organization manages risks and opportunities related to ESG factors. ESG reporting is a critical tool for stakeholders to evaluate a company's long-term value and sustainability practices. Globally, ESG has become the leading sustainable investment strategy, gaining significant traction among investors who seek investments that are both responsible and financially sound. ESG investment is often seen as a synergy between profitability and sustainability, rather than a trade-off between the two.

2.1.3 The Importance of Environmental Sustainability in ESG Reporting

Environmental sustainability involves the preservation and protection of the natural environment while meeting present needs and conserving resources for future generations. It is influenced by a complex interplay of factors, including human-induced climate change, air and water pollution, soil contamination, resource overexploitation, loss of biodiversity, and unsustainable consumption patterns. Achieving this form of sustainability requires a collective commitment, encompassing efforts to reduce greenhouse gas emissions, advocate for renewable energy adoption, implement policies for biodiversity conservation, embrace sustainable practices, raise awareness, engage communities, and apply the principles of a circular economy. Moreover, global initiatives, such as the Carbon Border Adjustment Mechanism (CBAM), the European Union's Corporate Sustainability Reporting Directive (CSRD), the RE100 commitment, Cradle to Cradle Certification, LEED standards, circular economy practices, and Sustainable Development Goals (SDGs), play vital roles in advancing environmental sustainability.

In the context of ESG investments, environmental sustainability assumes a new role, representing a significant role in the financial world. Investors now recognize the importance of incorporating environmental considerations into investment decisions alongside traditional financial criteria. The "E" in ESG investments emphasizes the significance of environmentally responsible practices and sustainability efforts aimed at minimizing environmental impacts. This approach entails a comprehensive evaluation of various environmental factors within a company's operations, including aspects such as carbon emissions, energy consumption, waste generation, water usage, biodiversity preservation efforts, and pollution control measures. Strategies and objectives such as reducing the carbon footprint, embracing circular economy practices, and complying with environmental regulations are effective means of addressing these considerations. The goal of integrating environmental factors into ESG investing is to achieve financial returns while prioritizing sustainability and environmental responsibility.

ESG investing relies significantly on environmental sustainability. It recognizes the profound impact of environmental concerns on a company's financial performance, reputation, and long-term viability. The environmental (E) pillar explores a crucial question: *How does the company's behavior impact the environment?* (Dragomir, 2020). This E pillar serves as a requirement in ESG reporting, providing a platform for companies to demonstrate their commitment to transparency, accountability, and a proactive approach to addressing risks associated with unsustainable practices.

Environmental ESG reporting is a specialized focus that considers a company's environmental practices, performance, and impact concerning critical factors such as climate change, resource conservation, and pollution mitigation. This strategic emphasis within the ESG framework not only positions companies favorably for resilience and success in today's rapidly changing world but also plays an important role in providing stakeholders with transparent information about a company's environmental initiatives. From setting carbon reduction targets to implementing energy efficiency measures and waste reduction efforts, ESG reporting, particularly on the environmental front, ensures adherence to environmental regulations and fosters transparent reporting practices. Moreover, environmental sustainability, as a core pillar of ESG, serves two purposes: informing investors and stakeholders about an organization's environmental stewardship and aligning business operations with the imperative of preserving the planet for future generations. Within the multifaceted framework of ESG, the environmental aspect conveys a narrative of conscientious corporate responsibility and a dedicated commitment to a sustainable future.

2.1.4 Integration of Environmental Factors into Investment Strategies

While sustainable finance is gaining increasing acceptance within the financial industry and among policymakers, substantial work remains to be done. The urgency is emphasized by a caution from a 2018 report by the Intergovernmental Panel on Climate Change (IPCC): if greenhouse gas (GHG) emissions persist at their current rate, the Earth's atmosphere is projected to warm by 1.5 °C above pre-industrial levels by 2030. This impending reality significantly exceeds the target agreed upon by 184 countries in the 2015 Paris Agreement, which aims to restrict the global average temperature increase to "well below 2°C above pre-industrial levels" and strives to limit it to 1.5 °C. In addition to the potential disastrous consequences for agriculture, coastlines, vital ecosystems, and poverty, this trajectory comes with a staggering global price tag, estimated between US$54 trillion and US$69 trillion, as highlighted by the (Task Force on Climate-related Financial Disclosures (TCFD), 2020).

Environmental ESG investing shapes investment decisions based on environmental criteria, while ESG reporting facilitates communication and accountability, allowing stakeholders to evaluate a company's environmental stewardship. Key environmental factors considered encompass climate change, carbon footprint, resource utilization, biodiversity, and pollution. A holistic view of environmental factors and detailed concerns of investors in the economic sphere are available in Table 2.1.

Before incorporating environmental factors into their investment strategies, companies must gain a comprehensive understanding of their existing conditions. This requires an in-depth comprehension of their business processes, which encompass inputs, business activities, outputs, and outcomes (The Committee of Sponsoring Organizations of the Treadway Commission (COSO) and World Business Council for Sustainable Development (WBCSD), 2018). *Inputs* encompass the essential resources upon which the business relies, including ecosystem services, raw materials, natural resources, labor, and water resources. It is imperative for companies to consider the stocks and flows of these crucial capitals, as they significantly influence the robustness and resilience of the business model. When assessing *business activities*, companies should critically evaluate the adaptability of their organizational design to accommodate change. This evaluation ensures that the company is well-prepared for shifts in environmental factors. In the context of *outputs*, it is crucial for companies to consider the impacts and potential consequences stemming from the products and waste generated throughout the value chain. Furthermore, the

Table 2.1 Environmental factors: an investor's perspective

Environmental factors	Definition	Investor's perspective
Climate change	Alterations of weather patterns caused by both natural processes (e.g., volcanic eruptions and solar radiation) and human activities (e.g., burning fossil fuels and deforestation) (Forster et al., 2007)	Potential risks that raise significant concerns about financial implications, including physical risks and transition risks (Climate Disclosure Standards Board (CDSB) and Sustainability Accounting Standards Board (SASB), 2019)
Carbon emissions	One of the greenhouse gases released primarily through human activities (e.g., burning fossil fuels, industrial processes, and deforestation), contributing significantly to the ongoing issue of climate change by trapping heat in the atmosphere and leading to global warming (Forster et al., 2007)	A key focus of ESG analysis for investors to understand the environmental impact and sustainability of their investments through metrics like carbon footprint and carbon intensity, which evaluate the emissions efficiency of companies within their portfolios (Principles for Responsible Investment (PRI), 2015)
Pollution	The effects of, for example, chemicals, particulate matter, or excessive heat on ecosystems, living organisms, and human health, causing pollution-related diseases, water contamination, biodiversity loss, and climate change (United Nations environmental programme (UNEP, 2017))	The environmental and financial risks associated with the release of harmful substances into the ecosystem by companies or industries in which they invest include of regulatory fines, legal liabilities, reputational damage, and increased operational costs
Resource efficiency	Optimizing the use of Earth's limited resources in a sustainable manner while minimizing environmental impacts to promote the responsible and efficient use of resources across various sectors of society, including industry, agriculture, and households (LIFE Fit for REACH and the LIFE Programme of the European Union, 2022)	The strategic allocation and management of resources, including energy, materials, and water, within a company or investment portfolio to maximize financial returns while minimizing waste and environmental impact, to mitigate risks associated with resource scarcity, regulatory changes, and shifts in consumer preferences
Biodiversity	The variety of life on earth at all levels of biological organization is essential for the health and resilience of ecosystems. It provides a range of ecosystem services, such as pollination, water purification, and climate regulation, upon which human societies also depend (the Secretariat of the Convention on Biological Diversity (SCBD), 2023)	Biodiversity loss poses significant risks, including regulatory, reputational, and operational risks, which can affect the value of investments. In addition, biodiversity can offer opportunities, such as investments in sustainable agriculture and eco-tourism (Climate Disclosure Standards Board (CDSB) and Sustainability Accounting Standards Board (SASB), 2019)

outcomes and contributions of the business, including its environmental impacts and the sustainability of its license to operate, must be rigorously assessed. This examination is essential for gaining a comprehensive understanding of the company's current standing and situation.

An example highlighting the importance of understanding a business's current environmental sustainability condition is the case of Coca-Cola in 1993 when they established a bottling facility in a water-scarce region of India. The company conducted an analysis to assess the impacts and interdependencies of its operations, realizing that the production of beverages depended on the availability of water within the host country. Consequently, local authorities decided to close Coca-Cola's factories. This decision stemmed from the local watershed's inability to meet both the community's water needs and the manufacturing operations (The Committee of Sponsoring Organizations of the Treadway Commission (COSO) and World Business Council for Sustainable Development (WBCSD), 2018).

Conducting a risk and opportunity assessment of environmental factors is now a standard practice in investment due diligence. This assessment helps investors identify and mitigate risks while seizing opportunities in the evolving landscape of sustainable finance, thereby influencing asset allocation and portfolio construction. The assessment includes environmental risks (e.g., natural catastrophes, climate change impacts, increasing volatility of weather, regulatory compliance) and opportunities (e.g., clean technologies, green innovation), as well as materiality identification and assessment, guiding asset allocation decisions.

2.2 Investors' Awareness of Climate Risk

2.2.1 Climate Risk Assessment

Climate change serves as one of the foremost drivers for the global sustainability movement. It can be defined as any alteration in climate patterns over time, whether due to natural fluctuations or as a consequence of human activities, as stated by the Intergovernmental Panel on Climate Change (IPCC). According to the United Nations Framework Convention on Climate Change (UNFCCC) (Forster et al., 2007), climate change can also be defined as "a change in climate attributed directly or indirectly to human activity that alters the composition of the global atmosphere and that is in addition to natural climate variability observed over comparable time periods." Climate change is a significant ESG factor, and it falls under the "E" (environmental) category.

ESG analysis evaluates a company's performance and risk exposure across a wide spectrum of nonfinancial considerations. In this context, investors should recognize climate change as a risk factor. This risk, referred to as *climate risk*, relates to the potential negative impacts and vulnerabilities faced by individuals, communities, businesses, and ecosystems due to the changing climate and related environmental factors. These risks can result in significant consequences, including damage to infrastructure, economic disruptions, threats to food and water security, population displacement, and harm to natural ecosystems. According to the recommendations of the TCFD, climate risks can be classified into two main categories: physical risk and transition risk.

Physical risk is related to the tangible and often immediate or long-term impacts of climate change, encompassing phenomena like extreme weather events, temperature increases, rising sea levels, and other physical alterations in the climate system. Instances of physical climate risks include heightened occurrences and intensity of natural disasters (e.g., hurricanes, floods, wildfires), damage to property and infrastructure due to sea-level rise and coastal erosion, decreased agricultural productivity stemming from shifting precipitation patterns and temperature extremes, as well as health hazards arising from heatwaves, air pollution, and the proliferation of vector-borne diseases. Furthermore, the comprehensive assessment of physical risk related to climate change strengthens the environmental sustainability perspective. Investors carefully evaluate the vulnerability of their investments to physical impacts, such as extreme weather events, sea-level rise, and supply chain disruptions driven by climate-related factors. Companies demonstrating robust strategies for managing these physical risks are viewed more favorably, reflecting their heightened commitment to environmental sustainability.

Case: Hurricane Katrina and the Insurance Industry

Background: Hurricane Katrina, one of the deadliest and costliest hurricanes in US history, struck the Gulf Coast in August 2005. The hurricane caused widespread destruction, primarily affecting the city of New Orleans. While this case primarily focuses on the insurance industry's response, it emphasizes the broader issue of physical climate risk.

Physical risk:

1. *Extreme weather event*: Hurricane Katrina was a catastrophic Category 5 hurricane, resulting in extensive flooding, storm surges, and wind damage in the affected areas.
2. *Sea-level rise*: The Gulf Coast's low-lying geography and vulnerability to sea-level rise made the region more susceptible to storm surges and coastal flooding.

This case demonstrates the financial and operational consequences of physical climate risks on the insurance industry and the broader economy. It underscores the need for businesses, governments, and insurers to prepare for and adapt to the physical impacts of climate change. The insurance industry has played a significant role in both addressing and responding to these risks, as it has had to adjust its risk management strategies and pricing to account for the changing climate landscape.

The aftermath of Hurricane Katrina posed significant challenges to the insurance industry and highlighted the physical climate risks associated with extreme weather events. The case illustrates how these risks can lead to substantial financial losses and impact the availability and affordability of insurance coverage.

1. *Insurance industry losses*: The insurance industry faced enormous losses due to the extensive damage caused by Hurricane Katrina. It prompted one of the largest insurance payouts in history, totaling tens of billions of dollars. The losses resulted from property damage, business interruption, and other claims.
2. *Impact on insurance market*: Following the hurricane, insurance companies began reassessing their risk models, underwriting practices, and premium pricing. Insurers had to account for the increasing frequency and severity of extreme weather events caused by climate change. This led to tighter underwriting standards, higher premiums, and reduced availability of coverage for high-risk areas.

The lessons learned from Hurricane Katrina have influenced how the insurance industry approaches climate-related risks, emphasizing the importance of adaptation, resilience, and sustainable urban planning in the face of physical climate challenges.

Transition risk, which emerges from policies, regulations, and market dynamics aimed at addressing climate change, stems from the transition toward a low-carbon economy and the reduction of greenhouse gas emissions. This shift requires adjustments in policy, legal frameworks, technology, and market conditions to meet the prerequisites for climate change mitigation and adaptation. It brings about a range of impacts on businesses and investments, including regulatory changes like carbon pricing, carbon taxation, or emissions reduction goals, resulting in varying financial and reputational implications for the involved entities. Moreover, transition risks arise from changes in market demand for products with high carbon intensity, the possibility of investments becoming economically unviable due to climate-related policies, and the risk to a company's reputation if it has high carbon emissions.

It is crucial for investors to assess transition risk by considering how regulatory changes and market shifts may impact their investments, all while prioritizing environmental sustainability principles. Companies that do not proactively adapt to developments like carbon pricing, evolving regulations, or changing consumer preferences toward eco-friendly products are perceived as posing elevated risks. Consequently, investors scrutinize how companies navigate the transition to a low-carbon economy, emphasizing the significance of environmental sustainability in their strategies.

Case: Coal Industry Transition

Background: The coal industry has long been a major contributor to greenhouse gas emissions and is considered a high-carbon industry. As global efforts to combat climate change have intensified, the coal industry has faced significant transition climate risks.

Transition risks:

1. *Regulatory Risks*: Governments around the world have introduced stricter regulations aimed at reducing carbon emissions. These regulations include emissions targets, carbon pricing mechanisms, and environmental standards. In many cases, coal-fired power plants and coal mining operations have become subject to costly compliance measure.
2. *Economic Risks*: The transition to cleaner energy sources, such as natural gas, renewables, and energy efficiency measures, has resulted in decreased demand for coal. This has led to economic risks for companies and regions heavily dependent on the coal industry. Coal prices have declined, and coal companies have faced financial challenges, including bankruptcies.
3. *Technology Risks*: As the world shifts toward cleaner technologies, coal companies have faced risks associated with obsolete technology and the need to invest in cleaner, more efficient processes. The costs associated with adopting cleaner technologies and practices can be substantial.

The case of Peabody Energy, one of the world's largest coal companies, illustrates transition climate risks. Peabody Energy filed for Chapter 11 bankruptcy in 2016, citing a sharp decline in coal prices and demand. The company had invested in high-cost coal assets while failing to anticipate the shift to cleaner energy sources and stricter environmental regulations. The bankruptcy resulted in significant financial losses for investors and demonstrated the risks of failing to adapt to the changing energy landscape.

This case highlights how companies and industries with high carbon footprints can face transition climate risks when they fail to adapt to the global shift toward sustainability and reduced carbon emissions. Transition risks can result in *financial losses, stranded assets, and operational challenges*, making it crucial for businesses to proactively assess and manage these risks in a changing climate-conscious world.

Investors are actively assessing climate risks in the broader context of environmental sustainability, recognizing the profound influence of climate change on financial considerations. An important strategy for this assessment is the integration of ESG factors into their investment decisions, with a heightened focus on the environmental dimension. They scrutinize ESG data and scores to gain insights into how companies manage climate-related challenges and opportunities within the framework of environmental sustainability. High ESG scores signify a company's commitment to sustainable practices and its resilience in the face of climate risks, while lower scores may raise concerns about vulnerabilities to these risks. Nonetheless, it's essential to recognize that having a high ESG rating does not automatically equate to a genuinely sustainable company. For example, a company heavily involved in fossil fuel industries that effectively manages associated risks may achieve a commendable ESG rating. Genuine sustainability means avoiding harm to the environment and society. In other words, *a good ESG score primarily indicates effective internal risk management, not necessarily a positive external impact.*

In summary, investors take a multifaceted approach to assess climate risks. They incorporate ESG criteria, scenario analysis, transition risk assessment, and physical risk assessment into their evaluation. Throughout this comprehensive process, investors place paramount importance on environmental sustainability as a guiding factor in their investment decisions. These methodologies enable investors to make informed choices, support environmental sustainability practices, and align their investments with a sustainable, low-carbon future. Climate risk assessment is a process of evaluating and quantifying the potential risks and vulnerabilities associated with climate change and its impacts on various sectors, industries, and investments. This assessment helps organizations, investors, and governments understand the potential risks they face due to climate-related factors and formulate strategies to mitigate and adapt to these risks. Among various climate risk assessment methods, the two main methods, often employed by organizations, governments, and businesses to measure, report, and manage their environmental impact and climate-related risks, are climate-related disclosure and reporting and carbon footprint.

2.2.2 Companies' Climate-Related Disclosure and Reporting

Environmental factors can be integrated into corporate practices through ESG disclosure and reporting. Understanding the synergy among diverse ESG reporting frameworks is crucial for facilitating the effective voluntary disclosure of environmental factors. Exploring into the area of ESG reporting frameworks, standards, and associated guidelines available to organizations for the voluntary disclosure of climate impacts, risks, and other environmental aspects, such as water use and plastic waste, is invaluable. Although adherence to these reporting structures is not obligatory for voluntary reporting, they can provide valuable support. Some frameworks even provide specialized protocols for specific industries, while others are adaptable for use by both organizations and municipalities.

In the ESG reporting ecosystem, *three major components* are frameworks, standards, and protocols. Frameworks establish the foundational structure for sustainability reporting by guiding the identification, assessment, and communication of important sustainability issues. In contrast, standards offer precise requirements, metrics, and indicators to measure and report on sustainability topics, ensuring uniformity across diverse organizations and sectors. Protocols, as essential tools, and methodologies, facilitate the collection, calculation, and disclosure of sustainability data in adherence to chosen frameworks and standards. They provide specific guidance on data collection and reporting, addressing a range of sustainability aspects, including GHG emissions, water consumption, waste management, energy usage, and plastic waste generation. These protocols may either be universally applicable or tailored to specific frameworks and industries, furnishing practical instructions for precise and reliable sustainability reporting.

Standards Standards are more specific and detailed guidelines that build upon the principles outlined by frameworks. They define the precise requirements, metrics, and indicators that organizations should employ when reporting on specific sustainability topics. Standards facilitate comparability across organizations and sectors by providing a shared language and a set of metrics for evaluating and disclosing sustainability performance. The nature of standards can vary based on the framework and industry sector. For example, the GRI offers a set of universal standards that are applicable to all organizations, alongside topic-specific standards that address issues specific to various industries. In contrast, the SASB concentrates on industry-specific standards designed to capture financially material ESG issues for companies within particular sectors. Therefore, standards are great tools for helping companies to be transparent on environmental issues and help investors to identify environmentally responsible companies.

International Financial Reporting Standards (IFRS S1 & IFRS S2): These standards are designed to seamlessly incorporate the recommendations of the Task Force on Climate-Related Financial Disclosures (TCFD) into ESG reporting (Climate Disclosure Standards Board (CDSB), 2021). They offer guidance to organizations on disclosing both sustainability-related financial information and specific climate-related data, ensuring compliance with evolving reporting requirements.

Frameworks ESG frameworks, or ESG reporting frameworks, provide organizations with structured guidelines to identify, evaluate, and report on sustainability issues relevant to their activities. These frameworks establish the foundational structure for ESG reporting by guiding the identification, assessment, and communication of important sustainability issues by facilitating organizations in measuring their performance against industry peers and global best practices. They also enable organizations to communicate their progress to a range of stakeholders, including investors, regulators, customers, and employees. Well-recognized ESG frameworks include the CDP (previously the Carbon Disclosure Project), the Global Reporting Initiative (GRI), the Sustainability Accounting Standards Board (SASB), and the Task Force on Climate-Related Financial Disclosures (TCFD). While each framework has its distinct focus, they all share a common objective of promoting transparency, comparability, and accountability in ESG reporting. Therefore, they are valuable tools for integrating environmental factors in ESG investment strategies.

Protocols Protocols constitute specific tools, methodologies, or instructions that guide organizations in measuring, monitoring, and reporting their ESG performance in accordance with the chosen framework and standards. These protocols may either be independent of or integrated into the frameworks. Protocols provide comprehensive guidance on the systematic collection, calculation, and accurate disclosure of data. They can encompass diverse aspects of ESG reporting, including greenhouse gas (GHG) emission accounting, water consumption, waste management, energy use, and plastic waste generation. Some protocols, like the GHG Protocol, have universal applicability across different frameworks as they offer universally accepted methodologies for measuring and reporting GHG emissions. Others may be more tailored to a particular framework or industry, as exemplified by the GRI's set of protocols embedded within its standards. Additional information on standards, frameworks, and protocols are provided in Chap. 7.

GHG Protocol: As a universally applicable protocol, the Greenhouse Gas (GHG) Protocol provides essential guidelines for measuring and managing greenhouse gas emissions. It serves as a critical tool for organizations, enabling them to track their climate impact effectively while aligning with various reporting frameworks and standards.

CDP Questionnaires: These structured questionnaires offer organizations a framework for responding to the Carbon Disclosure Project's (CDP) annual disclosure requests. They play a significant role in facilitating the reporting of essential data related to carbon emissions, water usage, and forest-related information.

CDP-Water Protocol: The CDP-Water Protocol provides valuable guidance on reporting water-related risks, opportunities, and impacts. By adhering to this protocol, organizations can meet CDP's water disclosure requirements and enhance transparency regarding their water-related activities.

CDP-Forest Protocol: For organizations looking to report on forest-related risks, opportunities, and impacts, the CDP-Forest Protocol offers clear guidelines. It aids in compliance with CDP's forest disclosure requirements, ensuring that relevant information is accurately reported.

Beyond standards, frameworks, and protocols, the ESG reporting ecosystem comprises various interconnected components. *Ratings and rankings* play an important role in assessing and positioning organizations based on their sustainability performance. Ratings involve the evaluation and scoring of organizations according to specific sustainability criteria, while rankings provide a comparative list that reflects an organization's performance relative to its peers and industry standards. Notable examples of rating agencies and rankers include MSCI, Sustainalytics, S&P Global, Moody's, and Science-Based Targets. Concurrently, regulations, typically enforced by governmental or regulatory authorities, establish mandatory ESG reporting guidelines. Examples of regulatory authorities that mandate ESG data and performance disclosures include the Corporate Sustainability Reporting Directive (CSRD), Mandatory Climate Risk Reporting (alignment to TCFD), EU Taxonomy Regulation, and Sustainable Financial Disclosure Regulation (SFDR). In addition, global goals, exemplified by the United Nations (UN) Sustainable Development Goals (SDGs), define universally recognized targets and objectives, guiding sustainability initiatives on a global scale. These components collaboratively influence and reinforce each other, fostering a cohesive sustainability reporting framework that integrates diverse aspects of performance evaluation, mandatory compliance, and shared sustainability objectives.

The incorporation of standards, frameworks, and protocols into ESG reporting represents a dynamic and interconnected process, rather than a sequential one. Nonetheless, to gain a clearer perspective on their interrelations, one can consider the process through the following sequence:

1. *Standard Identification*: Initially, organizations determine the specific standards to apply. These standards can be either universally applicable or industry-specific, depending on the organization's needs. These standards are vital for identifying and reporting key sustainability issues relevant to the organization.
2. *Framework Selection*: After choosing the appropriate framework, organizations proceed to the next step, which involves organizations making the critical decision to select an appropriate ESG reporting framework. This selection depends on how well the chosen framework aligns with the organization's strategic objectives, meets stakeholder expectations, and suits the industry's unique context.
3. *Protocol Application*: In the final step, organizations implement the relevant protocols, using them as tools for the measurement, monitoring, and reporting of their sustainability performance in accordance with the selected framework and standards.

These components empower companies to not only communicate their environmental initiatives but also effectively manage climate-related risks and opportunities, fostering a more sustainable and accountable corporate environment. By cohesively integrating these components, organizations have the opportunity to elevate the creditability, transparency, and influence of their sustainability reporting, fostering a positive transformation toward a more sustainable future.

2.3 Climate Risk Management

Climate risk management is a crucial aspect of environmental sustainability, particularly in the context of investments. It is conducted after a comprehensive assessment of the potential impacts of climate change on business models, strategies, and financial planning across short-, medium-, and long-term horizons, referred to as "strategy," which is one of the core elements in *the TCFD recommendations*. In recent years, investors have increasingly recognized the potential adverse effects of climate change on their portfolios. Understanding and addressing these risks is essential for making informed investment decisions and promoting responsible financial practices.

Transparency and disclosure are integral elements of effective climate risk management, providing investors with essential insights into a company's climate risk exposure and mitigation strategies. Investors are increasingly reliant on corporations for comprehensive climate risk disclosure, enabling them to make informed investment choices and encourage more sustainable corporate practices. International frameworks such as the TCFD have emerged to standardize climate risk reporting, streamlining the assessment of investment risks and opportunities.

Nonetheless, inadequate information concerning an organization's climate-related risk management poses challenges for investors in understanding the entity's overall risk profile. This underscores the vital role of TCFD recommendations, which promote universal climate-related risk management disclosure across all companies. Such disclosure should encompass the integration of climate risk practices into the organization's broader enterprise risk management framework. As a result, companies are progressively adopting the TCFD framework as an effective instrument for disclosing financial information related to climate considerations, with a particular focus on robust climate risk assessment and the seamless integration of climate-related factors into their financial reporting practices (Fig. 2.6).

The TCFD aims to enhance the transparency and reliability of climate-related financial disclosures. By placing a strong emphasis on governance, strategy, risk management, metrics, and targets, TCFD ensures the comprehensive integration of climate considerations throughout organizations. This approach fosters transparency and consistency while encouraging the use of *climate scenario analysis* to assess an organization's strategic resilience in the face of climate change and decarbonization efforts. This analysis aids companies in understanding potential impacts and adapting their strategies across short-, medium-, and long-term climate change considerations.

Scenario analysis forms a fundamental element within ESG core considerations, primarily within the "strategy." The TCFD highlights its vital role in assessing climate-related risks' potential impacts on business, strategy, and finances, functioning as a valuable tool for forward-looking risk assessment amid the inherent uncertainties of climate change. These uncertainties are influenced by various factors, encompassing dynamic change drivers, underlying assumptions, scenario

Fig. 2.6 The core elements of the TCFD recommendations

logic, and intricate interactions among physical climate elements. The aim is to construct scenarios that capture historical trends and anticipate future developments. Each scenario examines distinct high-impact and high-uncertainty driving forces, ensuring diversity and challenge. For instance, to evaluate transition risks, a company can create scenarios, like one where it meets the climate goals of the Paris Agreement. This involves exploring the necessary factors for that outcome. Other scenarios can consider different emission levels and various assumptions about the factors required for those outcomes.

Since the release of its final recommendations in June 2017, the TCFD has actively monitored and faced challenges related to implementing its guidelines for climate-related financial disclosures. A TCFD survey revealed that 75% of companies found it somewhat or very challenging to implement this recommendation. Furthermore, a notable proportion of these companies indicated that they lack processes for identifying, assessing, or managing climate-related risks. Therefore, this section will focus on *the TCFD's Risk Management recommendation*. The TCFD's disclosure recommendations are listed in Table 2.2.

In October 2020, the TCFD released guidance centered on risk management integration and disclosure. This guidance serves as a valuable resource by utilizing the Committee of Sponsoring Organizations of the Treadway Commission's (COSO's) enterprise risk management (ERM) framework as a basis for addressing risk management topics. The COSO's ERM framework offers a potential approach for boards and management to identify, manage, and align risks within predefined risk tolerance levels, ultimately supporting the achievement of objectives. It outlines considerations for interconnecting risk with strategic planning and day-to-day operations while fostering a corporate culture that embeds risk awareness and incorporates risk into performance management practices.

The COSO framework comprises five key components: governance and culture, strategy and objective setting, performance, review and revision, and information, communication, and reporting. While all these components contribute to effective risk management, this guidance focuses primarily on *the "Performance" component, aligning with the TCFD's Risk Management recommendation*. The "performance" component involves a process for identifying and assessing risks that have the potential to impact the achievement of strategic goals and business objectives. These risks are then prioritized according to their severity, considering the organization's risk tolerance. Subsequently, the organization selects appropriate responses to mitigate these risks. It's worth noting that, although this guidance incorporates concepts from COSO's ERM framework, it can be effectively used in conjunction with other risk management frameworks such as ISO 31000 or an organization's specific risk management processes.

The integration of climate-related risks into a company's existing risk management involves common processes: identifying risk, assessing and prioritizing risk, and implementing risk response (referred to differently but consistently performed). *Identifying risk* involves identifying new, emerging, or changing risks that may affect a company's business objectives. *Assessing and prioritizing risk* is about assessing risks to understand their impact on business objectives and prioritizing

Table 2.2 TCFD's 11 disclosure recommendations (TCFD, 2020)

TCFD's core elements	Disclosure recommendations
Governance The governance framework regarding climate-related risks and opportunities within the organization, involving the establishment of transparent board oversight and managerial responsibility for the management of climate-related risks and opportunities	1. Describe the board's management of climate-related risks and opportunities 2. Describe the management's involvement in assessing and managing climate-related risks and opportunities
Strategy Analyzing how climate change currently and potentially impacts the business model, strategy, and financial planning within short-, medium-, and long-term perspectives while also considering different potential scenarios. Assessing both the immediate and potential outcomes of climate-related risks and opportunities on the organization's businesses, strategy, and financial planning	3. Describe the climate-related risks and opportunities recognized by the organization in the short, medium, and long term 4. Outline the effects of climate-related risks and opportunities on the organization's operations, strategy, and financial planning 5. Describe the resilience of the organization's strategy, accounting for various climate-related scenarios, including those with a 2 °C or lower temperature increase
Risk assessment The procedures employed by the organization to identify, assess, and manage climate-related risks. Utilizing the suitable processes for identifying, evaluating, and managing climate-related risks	6. Describe the processes the organization employs to identify and assess climate-related risks 7. Describe the processes the organization utilizes for managing climate-related risks 8. Outline the integration of the processes for identifying, assessing, and managing climate-related risks into the organization's broader risk management framework
Metrics and targets The metrics and objectives employed to appraise and handle pertinent climate-related risks and opportunities. Employing the suitable internal metrics and objectives throughout the entire business model to evaluate and oversee pertinent climate-related risks and opportunities	9. Reveal the metrics employed by the organization to evaluate climate-related risks and opportunities in accordance with its strategy and risk management procedures 10. Share information regarding greenhouse gas (GHG) emissions, encompassing scope 1, scope 2, and, if applicable, scope 3 emissions, as well as the associated risks 11. Outline the objectives the organization employs to handle climate-related risks and opportunities, along with its performance in meeting these objectives

them to inform decisions on how to respond. Also, *implementing risk response* (managing risk) encompasses actions such as accepting, avoiding, mitigating, pursuing, or sharing risks based on the outcomes of risk prioritization.

The TCFD's guidance for integrating climate-related risks into current procedures is structured as a collection of foundational, preliminary actions, aiming to assist businesses in recognizing significant factors for integration. Moreover, the TCFD considers these initial actions, which are briefly outlined below, as iterative rather than strictly sequential:

- *Step 1* Ensure a comprehensive understanding of climate change concepts and its potential consequences is widespread throughout the organization.
- *Step 2* Identify the specific risk management processes and components that may require adjustments for the integration of climate-related risk, along with the functions and departments responsible for these processes and components.
- *Step 3* Incorporate climate-related risks in the existing risk taxonomy and risk inventory employed by the company. This includes aligning climate-related risks with pre-existing risk categories and types.
- *Step 4* Modify the existing risk management processes and essential components based on insights gained in the previous steps and the characteristics of climate-related risk.

Companies can effectively communicate issues related to climate-related risks and achieve growth, profitability, and resilient business operations through six essential steps:

1. *Where are you now?* Evaluate your company's current position in the industry and define your desired position.
2. *What is material?* Engage with stakeholders to gain insights into what is considered significant by assessing both the sustainability impacts of your company and its broader effects on society and the environment.
3. *Where is your direction?* Establish and define your company's objectives and governance strategies.
4. *What do you need to report and to whom?* Select the most suitable reporting framework to meet stakeholder expectations and clearly convey narratives about value creation.
5. *How will you prepare?* Prepare for disclosure by identifying and addressing data collection gaps and allocating the necessary resources for report production.
6. *How will you communicate?* Adhere to framework guidelines while reporting and effectively communicate with both internal and external stakeholders.

On the other hand, investors employ a variety of strategies to mitigate climate-related risks in their investments. Diversification is a common approach, spreading investments across various sectors and asset classes to reduce exposure to climate-sensitive industries. Some investors actively engage with companies in their portfolios to encourage sustainable practices and disclosures. Others opt for green investments, supporting ventures with a strong environmental focus. The integration of ESG criteria into investment decisions is another widely used strategy, helping investors identify companies with strong environmental practices. Within this framework, typically indicators encompass Carbon Price Integration, GHG Reduction Programs within the supply chain, Green Procurement Policies, GHG Emissions Policies related to the supply chain, Carbon Leadership Talent, Carbon Offsetting Programs, and Green Outsourced Logistics Initiatives.

According to the Sustainability Reporting Guide for Listed Companies by the Stock Exchange of Thailand (SET), the ESG Metrics section outlines five categories within the environmental dimension of ESG, which can be further categorized into core indicators and recommended indicators. These indicators encompass various aspects of a company's environmental policies, plans, and performance, specifically focusing on the efficiency of energy and resource utilization. A comprehensive list of these indicators is presented in Table 2.3. Rationale and Reporting Approach can be found in the SET report's appendix (The Stock Exchange of Thailand (SET), 2022).

In conclusion, climate risk management is a fundamental component of responsible investing in an era of climate change. Investors must be vigilant in identifying, assessing, and addressing climate-related risks to protect their portfolios and

Table 2.3 Environmental indicators in environmental dimension

Environmental dimension	Core (C) indicators	Recommended (R) indicators
E1 Environmental Policy and Compliance Standards GRI 103: Management Approach	E1.1C Environmental management policy and guidelines E1.2C Number of cases or incidents of legal violations or negative environmental impacts, with explanations of mitigation measures	E1.3R Value of damages or fines received from legal violations or negative environmental impacts E1.4R Compliance with international energy management principles and standards E1.5R Compliance with international water management principles and standards E1.6R Compliance with international waste management principles and standards
E2 Energy Management GRI 302: Energy SDG 7: Affordable and Clean Energy	E2.1C Energy management plan E2.2C Energy consumption (electricity/ fuel) E2.3C Renewable energy consumption	E2.4R Energy management target E2.5R Energy intensity
E3 Water Management GRI 303: Water and Effluents SDG 6: Clean Water and Sanitation	E3.1C Water management plan E3.2C Volume of water consumption	E3.3R Water use target E3.4R Water intensity E3.5R Percentage of wastewater treated before discharge

(continued)

Table 2.3 (continued)

Environmental dimension	Core (C) indicators	Recommended (R) indicators
E4 Waste Management GRI 306: Waste SDG 12: Responsible Consumption and Production	E4.1C Waste management plan E4.2C Volume of waste generated	E4.3R Waste management target E4.4R Volume of waste that is reused and/or recycled
E5 Greenhouse Gas Management GRI 305: Emissions SDG 13: Climate Action	E5.1C Greenhouse gas management plan E5.2C Scope 1 and 2 greenhouse gas emissions E5.3C External verification of greenhouse gas emissions data	E5.4R Greenhouse gas reduction target E5.5R Total greenhouse gas emissions (Scope 1, 2 and 3) E5.6R Carbon intensity

contribute to a more sustainable and resilient financial landscape. The strategies they employ, from diversification and engagement to ESG integration and green investments, play an important role in managing these risks. Moreover, the growing emphasis on climate risk disclosure and reporting is enhancing transparency and empowering investors to make well-informed decisions. As the understanding of climate risk continues to evolve, investors' commitment to proactive climate risk management is a vital step toward achieving a sustainable and responsible financial future.

2.4 Summary

Environmental factors play an important role in shaping investment strategies within the context of ESG investing, particularly in managing climate-related risks. As the financial impact of climate change becomes more pronounced, integrating environmental considerations into investment decisions has become essential. Tools such as scenario analysis and frameworks like the TCFD provide critical guidance for assessing the resilience of investments under various climate scenarios. These frameworks help investors identify, assess, and mitigate climate-related risks by promoting transparency and standardized reporting.

By embedding climate risks into existing enterprise risk management systems, companies can better align their business strategies with long-term environmental goals. This process involves integrating environmental factors into risk identification, prioritization, and response strategies. Investors also utilize various methods to manage these risks, including diversification, engagement with companies, and green investments. ESG criteria, including carbon pricing and greenhouse gas reduction targets, are becoming increasingly important in assessing investment opportunities. As the demand for robust environmental metrics grows, frameworks such as CDP, GRI, SASB, and TCFD play a significant role in shaping how companies disclose their environmental impact. Proactively addressing environmental risks not only enhances portfolio resilience but also aligns financial outcomes with sustainability objectives, demonstrating the importance of environmental factors in ESG investing.

2.5 Discussion Questions

1. How can investors effectively integrate climate-related risks into their existing risk management frameworks, and what are the key challenges they might face in doing so?
2. In what ways does the Task Force on Climate-related Financial Disclosures (TCFD) enhance transparency and standardization in climate risk reporting, and how can its recommendations influence corporate strategies?
3. How can scenario analysis be used as a tool to assess the long-term resilience of investments under varying climate scenarios, and what are the potential limitations of this approach?
4. What are some key strategies that investors can adopt to mitigate environmental risks in their portfolios, and how do these strategies align with broader sustainability goals?
5. How do global initiatives like the Paris Agreement and the Sustainable Development Goals (SDGs) influence ESG investing, and why is it important for investors to consider these frameworks when assessing climate-related risks?

References

Brundtland, G. H. (1987). *Our common future: Report of the world commission on environment and development*. Retrieved from Geneva: https://sustainabledevelopment.un.org/content/documents/5987our-common-future.pdf

Climate Disclosure Standards Board (CDSB). (2021). *TCFD good practice handbook* (2nd ed.). Climate Disclosure Standards Board (CDSB).

Climate Disclosure Standards Board (CDSB) and Sustainability Accounting Standards Board (SASB). (2019). *TCFD implementation guide*. Retrieved from https://www.tcfdhub.org/wp-content/uploads/2019/07/SASB_CDSB-TCFD-Implementation-Guide-A4-size-CDSB.pdf

Deloitte. (2022). *Working paper: Balancing your materiality assessment*. Retrieved from https://www2.deloitte.com/content/dam/Deloitte/nl/Documents/risk/deloitte-nl-risk-double-materiality.pdf

Dragomir, V. D. (2020). Theoretical Aspects of Environmental Strategy. In: Corporate Environmental Strategy. *SpringerBriefs in Applied Sciences and Technology*. Springer, Cham. https://doi.org/10.1007/978-3-030-29548-6_1

Elkington, J. (1997). The triple bottom line. *Environmental management: Readings and cases*, 2, 49–66.

Forster, P., Ramaswamy, V., Artaxo, P., Berntsen, T., Betts, R., Fahey, D. W., Haywood, J., Lean, J., Lowe, D. C., Myhre, G., Nganga, J., Prinn, R., Raga, G., Schulz, M., & Van Dorland, R. (2007). Climate change 2007: The physical science basis. In *Contribution of working group*

I to the fourth assessment report of the intergovernmental panel on climate change. Cambridge University Press.

International Integrated Reporting Council (IIRC). (2013). *The international integrated reporting <IR> framework*. Retrieved from IIRC: http://integratedreporting.org/wp-content/uploads/2013/12/13-12-08-THE-INTERNATIONAL-IR-FRAMEWORK-2-1.pdf

LIFE Fit for REACH and the LIFE Programme of the European Union. (2022). *Resource efficiency*. Retrieved from https://www.fitreach.eu/content/resource-efficiency#:~:text=Resource%20efficiency%20means%20using%20the,%2C%20recycling%20and%20eco%2Ddesign

Principles for Responsible Investment (PRI). (2015). *How measuring a portfolio carbon footprint can help*. Retrieved from https://www.unpri.org/climate-change/how-measuring-a-portfolio-carbon-footprint-can-assist-in-climate-risk-mitigation-and-reducing-emissions/608.article

Richardson, J., Steffen, W., Lucht, W., Bendtsen, J., Cornell, S. E., et al. (2023). Earth beyond six of nine Planetary Boundaries. *Science Advances*, 9, 37.

Stock Exchange of Thailand. (2014). Sustainability reporting guide for listed companies. https://media.set.or.th/set/Documents/2022/May/sd_report_2014_en.pdf

Task Force on Climate-related Financial Disclosures (TCFD). (2020). *Task Force on Climate-related Financial Disclosures: Guidance on risk management integration and disclosure*. Retrieved from https://www.fsb.org/wp-content/uploads/P291020-2.pdf

The Committee of Sponsoring Organizations of the Treadway Commission (COSO) and World Business Council for Sustainable Development (WBCSD). (2018). *Enterprise Risk Management: Applying enterprise risk management to environmental, social and governance-related risks*.

The Secretariat of the Convention on Biological Diversity (SCBD). (2023). *The convention on biological diversity*. Retrieved from https://www.cbd.int/convention/articles/?a=cbd-02

United Nations. (n.d.). Sustainable development: The 17 goals. United Nations. Retrieved March 5, 2025, from https://sdgs.un.org/goals

United Nations Economic and Social Commission for Asia and the Pacific (UNESCAP). (2018). *SEEA: Stocks and flows*. Retrieved from https://www.unescap.org/

United Nations Environmental Programme (UNEP) (2017). *Towards a pollution-free planet*. Retrieved from https://wedocs.unep.org/bitstream/handle/20.500.11822/21213/Towards_a_pollution_free_planet_advance%20version.pdf?sequence=2&isAllowed=y

WBCSD. (2017). *Sustainability and enterprise risk management: The first step towards integration*. Retrieved from https://www.wbcsd.org/contentwbc/download/2548/31131/1

Chapter 3
Environmental Performance Measuring

Kullapa Soratana

Abstract Environmental performance measurement is a key driver of sustainability in industries, and life cycle assessment (LCA) plays an important role in evaluating the environmental impacts of products and services throughout their life cycle. LCA consists of four primary steps—goal and scope definition, life cycle inventory analysis, life cycle impact assessment, and interpretation and improvement of the results—each contributing to identifying environmental hotspots, such as energy-intensive processes in the production of photovoltaic panels or the significant water usage in textile manufacturing. These steps help guide decision-making by highlighting areas where improvements can be made. While LCA provides transparency and comparability in assessing environmental impacts, it has limitations, particularly in capturing supply chain volatility, such as fluctuating material prices or the unpredictability of raw material availability. LCA often overlooks broader societal consequences, such as labor conditions or local community impacts, which are crucial for a comprehensive sustainability assessment. The integration of LCA into environmental, social, and governance (ESG) reporting frameworks, such as the Global Reporting Initiative (GRI) and the Sustainability Accounting Standards Board (SASB), allows organizations to disclose environmental performance in a structured manner. This enhances transparency, aligns with sustainability goals, and fosters informed decision-making, especially in areas like supply chain optimization or resource conservation. However, challenges remain, such as data gaps in emerging technologies or the need for cross-sector collaboration to standardize processes. Despite these challenges, innovations in technologies like AI and blockchain offer potential solutions, streamlining data collection and improving the accuracy of LCA results. By incorporating LCA into ESG reporting, organizations can make data-driven decisions that reduce environmental impacts and contribute to a more sustainable future.

K. Soratana (✉)
Faculty of Logistics and Digital Supply Chain, Naresuan University, Muang, Phitsanulok, Thailand

M. Bednárová, K. Soratana (eds.), *Environmental, Social, and Governance (ESG) Investment and Reporting*, https://doi.org/10.1007/978-3-031-84235-1_3

3.1 Addressing Other Environmental Impacts in Addition to Climate Change

3.1.1 *Expanding Beyond Carbon Footprint*

In recent years, the world of ESG investing and corporate reporting has undergone significant transformation. While ESG frameworks have traditionally placed a significant emphasis on reducing carbon emissions, there is now a consensus that sustainability considerations should transcend the limits of a company's carbon footprint. This transformation in ESG's scope necessitates a more comprehensive assessment of a company's impact on the environment, society, and governance practices, signaling a shift toward a holistic sustainability approach that acknowledges the multifaceted challenges of our ever-changing world.

Although addressing the carbon footprint remains crucial within ESG, it is no longer the sole metric for measuring an organization's commitment to responsible practices. In this evolving paradigm, attention now extends to a holistic evaluation of a company's operations, with a spotlight on critical aspects like water usage, biodiversity preservation, and pollution control. These dimensions of ESG highlight the intrinsic connection between business and the environment and underscore the growing understanding that responsible practices must encompass a comprehensive range of responsibilities to address today's multifaceted sustainability challenges.

Frameworks for Water and Biodiversity Disclosures: GRI (Global Reporting Initiative), TNFD (Task Force on Nature-Related Financial Disclosures), and CDSB (Climate Disclosure Standards Board), are presented in this section. Each provides frameworks and guidelines for water-related and biodiversity-related disclosures. GRI's standards encompass GRI 303, focusing on water and effluents, and GRI 304, emphasizing biodiversity and ecosystem services. TNFD, a newer initiative, calls for disclosures related to dependencies and impacts on nature, including water and biodiversity. CDSB, while primarily climate-focused, indirectly addresses water and biodiversity by emphasizing their interconnectedness with climate change and the need for comprehensive environmental disclosures. These frameworks collectively promote transparency and accountability in reporting, aiding organizations in assessing and disclosing their water and biodiversity-related risks, opportunities, and impacts in the context of nonfinancial and financial reporting. The relationship between GRI, TNFD, and CDSB and convergence of accounting and sustainability standards are presented in Fig. 3.1.

- *The Global Reporting Initiative (GRI) standards*

The Global Reporting Initiative (GRI) standards consist of a structured framework that includes three main components, universal standards, sector standards, and topic standards, as illustrated in Fig. 3.2. *Universal standards* provide foundational principles for sustainability reporting, offering guidance on report content, boundaries, and data quality. *Sector standards* are industry-specific and tailored to sectors like mining or construction, ensuring that organizations in these sectors

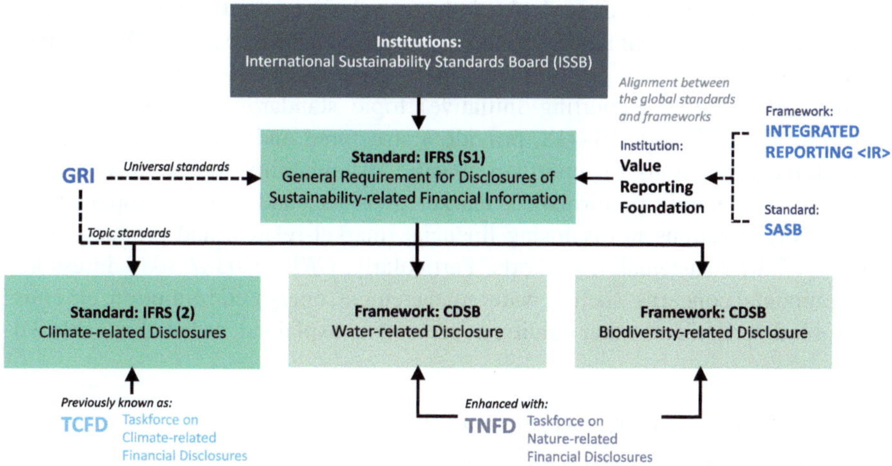

Fig. 3.1 The relationship between GRI, TNFD, and CDSB. (Author's elaboration)

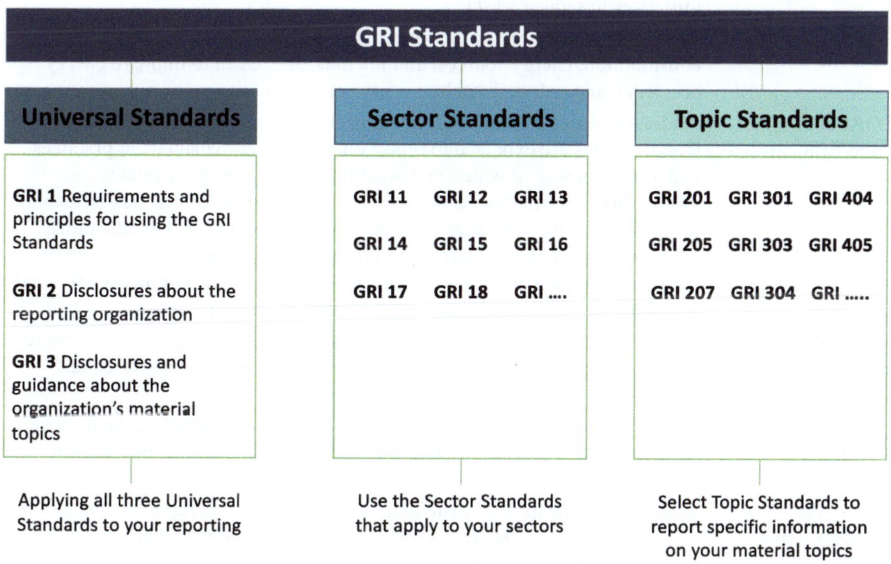

Fig. 3.2 GRI standards: universal, sector, and topic standards. (Adapted from the Global Sustainability Standards Board (GSSB), 2022)

address sector-specific sustainability issues. *Topic standards*, including those focusing on the environment, examine specific sustainability areas, offering in-depth guidance on reporting topics such as greenhouse gas emissions, biodiversity, and water management. This three-tiered structure allows organizations to build comprehensive and sector-specific sustainability reports by utilizing universal standards for consistency, sector standards for industry relevance, and topic standards for addressing specific sustainability concerns, fostering a holistic approach to sustainability reporting.

The GRI (Global Reporting Initiative) topic standards, encompassing GRI 201-208, 301-308, and 401-418, provide a structured and comprehensive framework for organizations to report on specific sustainability topics. *GRI 201-208* focuses on economic performance, market presence, and product responsibility, guiding organizations in disclosing financial, market-related, and product-related aspects of their sustainability efforts. Particularly, *GRI 301-GRI 308* address key environmental concerns, such as water management, energy consumption, air emissions, biodiversity, and water quality, as previously explained, as listed in Table 3.1.

Table 3.1 GRI topic standards, GRI 301 to GRI 308

Topic standards addressing environmental concerns	
GRI 301: Materials	Focuses on the efficient use of materials, emphasizing aspects like material consumption, energy efficiency, and waste generation. It requires organizations to report on their material usage, recycling efforts, and initiatives to reduce waste
GRI 302: Energy	Pertains to energy, requiring organizations to disclose their energy consumption, energy sources, and the measures taken to improve energy efficiency and reduce greenhouse gas emissions
GRI 303: Water and Effluents	Addresses water-related issues, emphasizing responsible water management. Organizations are expected to report on their water management practices, water use, sources of water, and potential impacts on water quality and availability. The standard also addresses effluents, including the disclosure of discharges and their associated environmental impacts. Organizations using this standard are expected to provide a comprehensive overview of their water-related and effluent-related activities, ensuring transparency and accountability in these areas of sustainability reporting
GRI 304: Biodiversity	Focuses on biodiversity and ecosystem services, requiring organizations to report on their impacts on ecosystems, habitat conservation efforts, and initiatives to restore or protect biodiversity
GRI 305: Emissions	Pertains to greenhouse gas emissions. It requires organizations to disclose their emissions, sources of emissions, and efforts to reduce their carbon footprint
GRI 306: Effluents and Waste	Addresses effluents and waste management. It requires reporting on the release of pollutants into air and water, as well as the management of hazardous and non-hazardous waste
GRI 307: Waste	Pertains to an organization's management of waste, including waste generation, disposal, recycling efforts, and initiatives aimed at minimizing the environmental impact of waste. Organizations are required to report on their waste-related activities, goals, and progress, fostering transparency and accountability in waste management as a critical aspect of sustainability reporting. Thank you for your patience and clarification
GRI 308: Supplier Environmental Assessment	Focuses on supplier environmental assessment, requiring organizations to report on the environmental performance of their suppliers and the steps taken to address environmental risks within the supply chain

Additionally, *GRI 401-418* focus on social aspects, covering topics like employment, labor practices, human rights, society, and product responsibility, offering detailed guidance on reporting in these areas. These topic standards collectively enable organizations to provide detailed and transparent reports on their sustainability performance and impacts, allowing stakeholders to gain valuable insights into their specific sustainability-related efforts and their associated effects. More details on GRI can be found in Chap. 7.

- *The Climate Disclosure Standards Board (CDSB)*

The Climate Disclosure Standards Board (CDSB) offers a framework and technical guidance for water-related and biodiversity-related disclosures that will be valuable for companies until the International Sustainability Standards Board (ISSB) introduces its IFRS sustainability disclosure standards. Established in 2007, CDSB pioneered a framework for companies to report environmental and social data with the same consistency as financial information. This framework laid a foundation for the TCFD recommendations, outlining an approach to incorporate environmental and social information into mainstream reports like annual reports, 10-K filings, and integrated reports. CDSB's framework for reporting environmental and social data, including climate and water disclosures, along with its extensive resources, will continue to be relevant and applicable for companies until the ISSB, an IFRS standards body, releases its corresponding IFRS Sustainability Disclosure Standards on these subjects.

- *The Task Force on Nature-Related Financial Disclosure (TNFD)*

The Task Force on Nature-Related Financial Disclosure (TNFD), a global initiative published in September 2023 and now poised for market adoption, shares similarities with the TCFD framework. TNFD comprises four key recommendations, aligning with the TCFD's structure, encompassing governance, strategy, risk management, and metrics and targets. However, TNFD distinguishes itself by its commitment to aligning its disclosure framework with the Kunming-Montreal biodiversity agreement, which strives to conserve and manage 30% of terrestrial, inland water, coastal, and marine areas by 2030, with the primary objective of achieving a net positive impact on nature. It builds upon the Natural Capital Protocol and introduces a localized impact assessment, guiding companies in comprehending the intricacies of their impacts and dependencies on natural resources through impact and dependency mapping. An impact pathway defines how specific business activities induce changes in natural capital and their subsequent effects on stakeholders, while a dependency pathway illustrates how particular business activities rely on specific aspects of natural capital. TNFD's comprehensive recommendations and guidance empower organizations to transparently report and address evolving nature-related dependencies, impacts, risks, and opportunities, facilitating the integration of nature considerations into decision-making processes for businesses and finance. This initiative ultimately steers global financial flows toward nature-positive outcomes, addressing the growing importance of nature-related frameworks and regulations in today's business landscape.

The TNFD framework is at the heart of the initiative, offering recommendations for organizations to address and disclose nature-related risks and opportunities. The latest version, beta v 0.3, released in November 2022, maintains the core components introduced in beta v 0.1. These components provide essential nature-related concepts and definitions for risk assessment and disclosure, draft disclosure recommendations, and guidance for integrating nature-related considerations into strategies and risk management. Within the framework, the risk and opportunity assessment component, known as *Locate, Evaluate, Assess, and Prepare (LEAP)*, as listed in Table 3.2., advises companies to identify their interface with nature, evaluate their dependencies and impacts, assess material risks and opportunities, and prepare responses, all to be reported to investors. Beta v 0.2, launched in June 2022, introduced an approach for steering nature-related metrics and targets, categorizing them into assessment (for internal decision making) and disclosure (for TNFD-aligned reporting).

Businesses and financial entities should adhere to a four-step process for recognizing their relationship with the natural environment. These four steps involve:

1. *Identifying*: Recognizing the points where business or financial activities intersect with the natural world, understanding the specific ecosystems in those locations, and identifying the crucial environmental assets and ecosystem services relevant to the enterprise.
2. *Responding*: Adapting by reevaluating strategic plans and business strategies, establishing and committing to targeted outcomes, and reviewing governance, risk management, and capital allocation processes.
3. *Assessing*: Evaluating dependencies on and impacts to the natural environment, along with assessing potential risks and opportunities for the business or capital portfolio.
4. *Disclosure*: If required, providing transparent reporting of outcomes, in conjunction with core financial statements and emissions reports, to external capital providers.

3.1.2 Key Environmental Metrics: Water Usage and Biodiversity

In the face of the growing recognition of environmental risks to financial and societal stability, many regulators are now prioritizing environmental and climate-related disclosures to promote climate action and mitigate related risks. Companies are beginning to feel the financially material impacts of competition for water

Table 3.2 TNFD's risk and opportunity assessment guidance (LEAP)

Locate	Evaluate	Assess	Prepare
Interface with nature	Dependencies and impacts	Material risks and opportunities	To respond and report
Engagement with affected stakeholders			
L1 Business footprint **Where** are our direct assets and operations and our related value chain (upstream and downstream) activities?	**E1** ID of relevant environmental assets and ecosystem services **What** are our business processes and activities at each priority location? What environmental assets and ecosystem services do we have a dependency or impact on at each priority location?	**A1** Risk and opportunity ID **What** are the corresponding risks and opportunities for our business?	*Strategy & resource allocation*: **P1** Strategy and resource allocation **What** strategy and resource allocation decisions should be made as a result of this analysis?
L2 Nature interface **Which** biomes and ecosystems do these activities interface with? What is the current integrity and importance of the ecosystems at each location?	**E2** ID of dependencies and impacts **What** are our nature-related dependencies and impacts across our business at each priority location?	**A2** Existing risks mitigation and risk and opportunity management **What** existing risk mitigation and risk and opportunity management approaches are we already applying?	**P2** Performance measurement **What** will we disclose in line with the TNFD disclosure recommendations?
L3 Priority location identification At **which** locations is our organization interfacing with ecosystems assessed as being low integrity, high biodiversity importance, and/or areas of water stress?	**E3** Dependency analysis **What** is the size and scale of our dependencies on nature in each priority location?	**A3** Additional risk mitigation and risk and opportunity management **Which** risks and opportunities are material and should be disclosed in line with the TNFD disclosure recommendations?	*Disclosure actions*: **P3** Reporting **Where** and **how** do we present our nature-related disclosures?
L4 Sector identification **What** sectors, business units, value chains, or asset classes are interfacing with nature in these priority locations?	**E4** Impact analysis **What** is the size and scale of our nature impacts in each priority location?	**A4** Risk and opportunity materiality	**P4** Presentation

resources and ecosystem degradation, posing significant risks to financial and societal systems at large. This section explores the disclosure of water and biodiversity-related aspects, which are addressed in frameworks such as GRI and CDSB.

- *Water-Related Disclosure Framework*

Water is a finite resource that different stakeholders share and compete for to meet their diverse needs and objectives. Its importance extends beyond environmental sustainability, playing a vital role in the stability of socioeconomic systems, as underscored by global initiatives like the Sustainable Development Goals (SDGs) and the concept of Planetary Boundaries. Despite the existence of sustainability reporting organizations and their guidelines for corporate water reporting, there's a noticeable gap in the availability of specific, nonsectoral recommendations for mainstream reports, akin to the TCFD's guidance for climate-related financial risks and opportunities.

Water-related concerns currently top the list of global environmental and societal risks, with businesses increasingly facing with substantial financial consequences linked to these risks. Water resources hold pivotal positions in enterprise operations, serving as integral components in production processes and as recipients for wastewater disposal. As a result, water-related risks have the potential to inflict harm on an organization's operational efficiency, reputation, legal standing, and financial performance. Consequently, it's imperative to integrate these risks into financial statements and their accompanying notes.

Investors are responding by aggregating corporate water-related data and integrating it into their portfolio risk assessments, creating dedicated investment funds for water, and launching initiatives aimed at raising awareness of water-related risks and opportunities. Despite some strides made by sectors heavily reliant on water, such as beverage and mining, there remains a need for continued efforts to ensure that the reporting of material water-related financial issues in mainstream reports meets the standards of quality and detail necessary to support decision-making by investors and other stakeholders, mirroring the success of the TCFD recommendations in promoting climate-related financial disclosure.

Nonetheless, businesses have the capability to play a central role in mitigating water-related risks through actions within their operations and supply chains, as well as on a broader water basin level. Discharge practices can significantly influence the functionality of natural ecosystems and the socioeconomic well-being of local communities within these basins. As a response, corporate water strategies and policies should strive for improvements, not only by reducing pollution and enhancing operational water efficiency but also by adopting a water basin approach. This approach recognizes water as a shared resource and considers the specific local environmental, regulatory, and socioeconomic context through active engagement with other stakeholders whenever practical.

The GRI Topic Standard 303: *Water and Effluents 2018*, established by the Global Sustainability Standards Board (GSSB), is a comprehensive framework developed with input from various stakeholders and in alignment with authoritative intergovernmental guidelines. This standard serves as a crucial tool for

Table 3.3 Reporting organization requirements for GRI 303: Water and Effluents 2018

Categories	Disclosures	Requirements
Topic management disclosures	303-1 Interactions with water as a shared resource	(a) A description of how the organization interacts with water, including how and where water is withdrawn, consumed, and discharged, and the water-related impacts the organization has caused or contributed to or that are directly linked to its operations, products, or services by its business relationships (e.g., impacts caused by runoff) (b) A description of the approach used to identify water-related impacts, including the scope of assessments, their timeframe, and any tools or methodologies used (c) A description of how water-related impacts are addressed, including how the organization works with stakeholders to steward water as a shared resource, and how it engages with suppliers or customers with significant water-related impacts (d) An explanation of the process for setting any water-related goals and targets that are part of the organization's approach to managing water and effluents and how they relate to public policy and the local context of each area with water stress
	303-2 Management of water discharge-related impacts	(a) A description of any minimum standards set for the quality of effluent discharge and how these minimum standards were determined, including: (i) How standards for facilities operating in locations with no local discharge requirements were determined (ii) Any internally developed water quality standards or guidelines (iii) Any sector-specific standards considered (iv) Whether the profile of the receiving waterbody was considered
Topic disclosures	303-3 Water withdrawal	(a) Total water withdrawal from all areas in megaliters, and a breakdown of this total by the following sources, if applicable: (i) Surface water (ii) Groundwater (iii) Seawater (iv) Produced water (v) Third-party water (b) Total water withdrawal from all areas with water stress in megaliters, and a breakdown of this total by the following sources, if applicable: (i) Surface water (ii) Groundwater (iii) Seawater (iv) Produced water (v) Third-party water and a breakdown of this total by the withdrawal sources listed in i-iv. (c) A breakdown of total water withdrawal from each of the sources listed in Disclosures 303-3-a and 303-3-b in megaliters by the following categories: (i) Freshwater (≤1000 mg/L total dissolved solids) (ii) Other water (>1000 mg/L total dissolved solids) (d) Any contextual information necessary to understand how the data have been compiled, such as any standards, methodologies, and assumptions used *Compilation requirements* When compiling the information specified in Disclosure 303-3, the reporting organization shall use publicly available and credible tools and methodologies for assessing water stress in an area

(continued)

Table 3.3 (continued)

Categories	Disclosures	Requirements
	303-4 Water discharge	(a) Total water discharge to all areas in megaliters, and a breakdown of this total by the following types of destination, if applicable: (i) Surface water (ii) Groundwater (iii) Seawater (iv) Third-party water and the volume of this total sent for use to other organizations, if applicable (b) A breakdown of total water discharge to all areas in megaliters by the following categories: (i) Freshwater (≤1000 mg/L total dissolved solids) (ii) Other water (>1000 mg/L total dissolved solids) (c) Total water discharge to all areas with water stress in megaliters, and a breakdown of this total by the following categories: (i) Freshwater (≤1000 mg/L total dissolved solids) (ii) Other water (>1000 mg/L total dissolved solids) (d) Priority substances of concern for which discharges are treated, including: (i) How priority substances of concern were defined, and any international standard, authoritative list, or criteria used (ii) The approach for setting discharge limits for priority substances of concern (iii) Number of incidents of non-compliance with discharge limits (e) Any contextual information necessary to understand how the data have been compiled, such as any standards, methodologies, and assumptions used *Compilation requirements* When compiling the information specified in Disclosure 303-4, the reporting organization shall use publicly available and credible tools and methodologies for assessing water stress in an area
	303-5 Water consumption	(a) Total water consumption from all areas in megaliters (b) Total water consumption from all areas with water stress in megaliters (c) Change in water storage in megaliters, if water storage has been identified as having a significant water-related impact (d) Any contextual information necessary to understand how the data have been compiled, such as any standards, methodologies, and assumptions used, including whether the information is calculated, estimated, modeled, or sourced from direct measurements, and the approach taken for this, such as the use of any sector-specific factors

organizations to disclose information regarding their water-related impacts and the strategies they employ to manage these effects. It is applicable to organizations of all sizes, types, sectors, geographic locations, and levels of reporting experience. When an organization deems water and effluents as a material topic, it is mandatory to include specific disclosures in their reporting, as specified in GRI 3: Material Topics 2021 (refer to clause 1.1 in this standard). Furthermore, the organization must incorporate any relevant disclosures from the GRI 303: Water and Effluents 2018 Topic Standard, specifically, Disclosure 303-1 through Disclosure 303-5.

GRI 303: Water and Effluents 2018 is structured into two categories of disclosures: topic management disclosures and topic disclosures. The former comprises Disclosure 303-1, which pertains to interactions with water as a shared resource, and Disclosure 303-2, which deals with the management of water discharge-related impacts. The latter includes Disclosure 303-3 on water withdrawal, Disclosure 303-4 on water discharge, and Disclosure 303-5 on water consumption. To ensure consistency and clarity in reporting, organizations are required to present the information in Table 3.3. For detailed recommendations and guidance on reporting each disclosure, organizations can refer to GRI 303: Water and Effluents 2018 topic standards, accessible through the provided the Global Sustainability Standards Board (GSSB) (2023).

The CDSB framework application guidance for water-related disclosures (the Water Guidance) aims to expand upon the principles of the TCFD recommendations by addressing matters related to natural capital (Climate Disclosure Standards Board (CDSB), 2021b). This supplementary document is the second in a series dedicated to enhancing the quality of disclosures pertaining to significant water-related concerns. The main objective of the Water Guidance is twofold: (1) it is designed to aid organizations in the identification and evaluation of financial information related to water for inclusion in mainstream reports, covering both risks and opportunities; and (2) it aims to streamline the creation of high-quality disclosures that enable users of mainstream reports to effectively assess critical financial data associated with water. The Water Guidance has a comprehensive scope, encompassing various aspects of water-related information that should be integrated into mainstream reports. It places particular emphasis on water governance, policies, strategies, targets, risk assessment, sources of environmental impact, performance analysis, comparative assessments, and outlook. Additionally, it explores into the assessment of the quantity and quality of freshwater resources, including both surface and groundwater. The document also provides initial reporting elements to address issues related to biodiversity, as well as the decline and deterioration of aquatic ecosystems, whether freshwater or non-freshwater in nature.

The complex relationship between water and businesses is characterized by several crucial aspects that organizations must thoroughly grasp and incorporate into their approach when addressing water-related risks and opportunities. These considerations are fundamental for developing and implementing effective strategies and determining the relevant water-related information to include in mainstream reports. Organizations should pay particular attention to the following factors:

1. *Site-Specific Water Issues*: Water dependencies, impacts, risks, and opportunities are inherently tied to specific locations, whether within an organization's own operations or along its entire value chain. This entails considering a comprehensive spectrum of factors in a given location, encompassing not only the physical

state of water resources in terms of quantity, quality, and temporal variations but also extending to aspects like water-related infrastructure management, social conditions, economic factors, governance and regulatory frameworks, and geopolitical dimensions, especially in transboundary basins.
2. *Value Chain Assessment*: Recognizing the global nature of modern value chains, water dependencies, impacts, risks, and opportunities often have the greatest significance beyond the boundaries of an organization. Water issues can influence the entire lifecycle within the value chain, from the use of water in raw material sourcing to waste management. Consequently, it is imperative for companies to incorporate their value chains into the assessment of water-related dependencies, impacts, risks, and opportunities.
3. *Collaborative Stakeholder Engagement*: Effective water management and stewardship require collaborative actions and engagement with value chain stakeholders at the basin scale. Actions undertaken solely at the operational level, such as increasing water efficiency, may not yield meaningful improvements in the status of water resources if other users within the same basin are degrading these resources and ecosystems. This scenario underscores the importance of regulators implementing basin plans and effectively regulating water use and discharge.

Water-related issues can be broadly categorized into key areas, each presenting distinct challenges. Firstly, there's the concern of insufficient water supply, characterized by an inadequate quantity of water to meet users' needs, including issues related to water infrastructure, distribution, and access. Conversely, excessive water issues arise from an overwhelming surplus, often due to flooding or extreme weather conditions with intense precipitation. Additionally, water pollution poses a significant challenge, involving contamination or alteration that renders water unsuitable for its intended use. These categories represent a range of critical water-related challenges that demand careful management. In addition to these primary water-related concerns, it's essential to acknowledge the intricate interplay of various environmental and socioeconomic factors, such as climate patterns, land cover, and socioeconomic conditions, particularly in regions with limited water resources. As a result, the development of effective and resilient water management strategies hinges on a comprehensive understanding of these dynamic and interconnected systems.

Organizations may encounter a range of financial risks and opportunities concerning water, encompassing physical, reputational, policy, technological, and market-related factors. Physical risks are linked to the quantity and quality of available water, including concerns about contamination and limited access. These physical risks encompass the potential dangers arising from water resource impacts, environmental systems, and processes influenced by factors such as climate change. They involve an increased likelihood and severity of extreme water-related weather events, water contamination, sea-level rise, shifts in precipitation patterns, greater water stress, changes in ecosystems, and loss of biodiversity. Additionally, organizations may struggle with transition risks associated with future water security, influenced by factors like regulatory adjustments, changing market preferences, stakeholder perceptions, and technological advancements.

Addressing water-related risks can also unlock business opportunities and financial advantages, such as improved water efficiency, the development of new products and services, and the conservation and restoration of ecosystems through collaboration with stakeholders. It's essential to recognize that these risks may arise from an organization's specific operations or value chain, as well as from the broader context within which they operate. This context can be shaped by factors like mismanagement by other water users and socioeconomic conditions within the operational basin. Furthermore, water-related risks are interconnected with wider environmental concerns, including land use and climate change. The impact of climate change, in particular, is already apparent through an increase in occurrences of floods, droughts, severe storms, and rising sea levels. Understanding these interrelated risks is critical for strengthening business resilience and ensuring continuity in the face of potential future scenarios.

The water-related disclosure checklist provides a summary of recommendations on how to incorporate essential information concerning water in mainstream reports while adhering to the CDSB standards. It's important to note that the items on the checklist are not obligatory mandates but rather recommended disclosures. They are to be included in the mainstream report if they are deemed significant to the organization. Information from the checklist is summarized and presented in Table 3.4.

- *Biodiversity-Related Disclosure Framework*

Biodiversity holds significant importance for both business and society, as emphasized by international initiatives such as the Sustainable Development Goals (SDGs), Planetary Boundaries, and the Dasgupta Review. Biodiversity is not only vital for achieving specific SDGs related to life below water and life on land but also plays a fundamental role in addressing hunger, promoting good health, responsible consumption, and climate action. It underpins the sustainability of natural and socioeconomic systems.

From a business and societal perspective, biodiversity is a foundation, constituting the living component of natural capital and supporting essential ecosystem services. These services encompass various benefits such as timber, pollination, water regulation, and recreation, which enable human activities, including business operations. Biodiversity's influence on ecosystem functioning and productivity enhances the quality, quantity, and resilience of ecosystems, resulting in more robust service provision.

For businesses, focusing on final ecosystem services in assessments and reports enhances accuracy and helps avoid common inaccuracies like double counting. Despite its pivotal role, human activities, including those of businesses, are driving an alarming and accelerating loss of biodiversity. This loss poses significant risks for the private sector, and businesses can play a crucial role in mitigating these risks through their operations, supply chains, and corporate biodiversity strategies.

The primary drivers of biodiversity loss, including land-use change, resource exploitation, climate change, pollution, and invasive species, are intricately linked

Table 3.4 CDSB framework's application guidance for water-related disclosure checklist

REQ-01 Governance Disclosure shall describe the governance of environmental policies, strategy, and information Does the disclosure: Identify the person(s) or committee responsible for water policies, strategy, and information? Explain how water policies, strategy, and information are delegated to management and if there are specific roles or mechanisms in place in hotspot areas to tackle compliance with water-related regulatory landscape and engagement with stakeholders? Describe any systems for accountability and incentivization? Explain whether the governance mechanisms for water policies, strategies, and disclosure differ from other material concerns and, if so, why?	*REQ-04 Sources of environmental impact* Quantitative and qualitative results, together with the methodologies used to prepare them, shall be reported to reflect material sources of environmental impact Does the disclosure: Provide metrics and indicators for sources of material water impacts (at least withdrawals, consumption, discharge) using both absolute and normalized metrics? Explain the metrics used, including the methodologies, geographic details, and levels of uncertainty, and provide narrative to assist in understanding of the results? Categorize and disaggregate metrics, e.g., considering areas affected by different levels of water risks or different water sources and destinations, to support understanding and comparability?
REQ-02 Management's environmental policies, strategy, and targets Disclosures shall report management's environmental policies, strategy, and targets, including the indicators, plans and timelines used to assess performance Does the disclosure: Explain the material water-related dependencies and impacts of the organization with additional consideration of links to natural capital? Summarize the water policies and strategies and how they support or link to the organization's risks and opportunities and overall strategy? When applicable, explain whether and how water strategies, policies, and management are influenced by stakeholder engagement? Set out the contextual, science-based, and time-bound targets, timelines, and indicators for delivery of water policy and strategy with methods and baselines, as well as explain progress and/or the development of policies? Detail the resourcing of the delivery and management of water policies and strategies?	*REQ-05 Performance and comparative analysis* Disclosures shall include an analysis of the information disclosed in REQ-4 compared with any performance targets set and with results reported in previous periods Does the disclosure: Provide appropriate historical data to the results reported from REQ-04 for material water impacts to allow for useful comparison, including details on hotspot areas? Contextualize the performance with baselines, targets, and other criteria used to assess progress? Explain the major trends with reference to drivers of change under (e.g., water-related strategies or business developments) and/or outside (e.g., regulatory changes) the control of the organization?

(continued)

Table 3.4 (continued)

REQ-03 Risks and opportunities Disclosures shall explain the material current and anticipated environmental risks and opportunities affecting the organization Does the disclosure: Identify material water-related risks and opportunities by adopting a basin-scale and value chain approach and by considering different types of risks? Explain the implications of material water-related risks and opportunities on business and value chains, specifying geographic locations and time horizons in which the risks will materialize? Describe the systems and processes used for assessing, identifying, and monitoring water- related risks and opportunities, including whether they are integrated with existing risk management systems and processes?	*REQ-06 Outlook* Management shall summarize their conclusions about the effect of environmental impacts, risks, and opportunities on the organization's future performance and position Does the disclosure: Explain the likely effect of future water-related impacts, risks, and opportunities as well as of water strategy on company performance and resilience, taking into account of regulatory and market trends and environmental changes? Identify and explain the time horizons used for reporting on corporate outlook? Explain any techniques, such as scenario analysis, used to inform the outlook including the methods, scenarios, and assumptions used, and any shortcomings and uncertainties?

to business activities. However, businesses can also be part of the solution by adopting sustainable practices and investing in nature-positive projects to conserve and restore natural ecosystems and biodiversity.

The GRI Topic Standard 304: *Biodiversity 2016*, established by the Global Sustainability Standards Board (GSSB), is a comprehensive framework developed with input from various stakeholders and in alignment with authoritative intergovernmental guidelines. This standard serves as a crucial tool for organizations to disclose information regarding their biodiversity-related impacts and the strategies they employ to manage these effects. It is applicable to organizations of all sizes, types, sectors, geographic locations, and levels of reporting experience. When an organization deems biodiversity as a material topic, it is mandatory to include specific disclosures in their reporting, as instructed in Top Management Disclosure 3-3 of GRI 3: Material Topics 2021 (see clause 1.1 in this section). Furthermore, the organization must incorporate any relevant disclosures from the GRI 304: Biodiversity 2016 Topic Standard, specifically, Disclosure 304-1 through Disclosure 304-4.

GRI 304: Biodiversity 2016 is structured into two categories of disclosures, topic management disclosures and topic disclosures. The former comprises Disclosure 3-3 in GRI 3: Materials Topics 2021, which outline how it intends to implement its biodiversity management policy. A biodiversity strategy may encompass various aspects, including preventing, managing, and mitigating harm to natural habitats caused by the organization's actions. An illustration of this is the incorporation of biodiversity concerns into analytical instruments like environmental site impact assessments. The latter includes Disclosure 304-1 on operational sites owned, leased, managed in, or adjacent to, projected areas, and areas of high biodiversity value outside protected areas; Disclosure 304-2 on significant impacts of activities, products, and services on biodiversity; Disclosure 304-3 on habitats protected or restored; and Disclosure 304-4 IUCN Red List species and national conservation list species with habitats in areas affected by operations. To ensure consistency and clarity in reporting, organizations are required to present the information in Table 3.5. For detailed recommendations and guidance on reporting each disclosure, organizations can refer to GRI 304: Biodiversity 2016 topic standards, accessible through the Global Sustainability Standards Board (GSSB) (2022).

- *The CDSB Framework Application Guidance for Biodiversity-Related Disclosures*

CDSB has developed the Biodiversity Application Guidance, an extension of the CDSB Framework, to aid companies in disclosing material information regarding the risks and opportunities of biodiversity on their strategy, financial performance, and overall condition in mainstream reports, known as biodiversity-related financial disclosure (Climate Disclosure Standards Board (CDSB), 2021a). This guidance is specifically designed to complement the CDSB Framework, which focuses on reporting environmental and climate change information to investors, and it serves as a valuable resource for organizations seeking to integrate biodiversity-related disclosures into their reporting practices. The Biodiversity Application Guidance serves a pivotal purpose: it is designed to aid organizations in preparing high-quality disclosures, thereby enabling users of mainstream reports to evaluate material financial information related to biodiversity. In a concerted effort to ensure that investors have the requisite biodiversity-related data for informed capital allocation, this guidance seeks to play a significant role in propelling the shift toward a sustainable and resilient economy. The target audience for this guidance encompasses a broad spectrum, including individual companies, corporate groups, as well as those with responsibilities in financial, governance, and sustainability reporting.

The relationship between businesses and biodiversity is characterized by dependencies and impacts. Biodiversity dependencies signify an organization's reliance on biological resources and interactions with ecosystem processes like pollination and water filtration. In contrast, biodiversity impacts encompass alterations in ecosystems and species diversity resulting from business actions, with outcomes that can be either positive, such as nature restoration, or more frequently negative, leading to potential biodiversity loss. These impacts can manifest in direct ways, such as immediate consequences of actions like land clearing, or indirectly through causal links, like greenhouse gas emissions contributing to climate change. They can also accumulate due to multiple factors, including various actors, background pressures, and trends. Importantly, these impacts are interconnected with dependencies through feedback loops, wherein a business may depend on a specific species but contribute to its population decline through unsustainable practices, affecting productivity and costs.

Definitions

Biodiversity impact: a change in the diversity of ecosystems and species can occur as a result of business activities. These changes in the condition and size of ecosystems, as well as the habitats and populations of species, serve as indicators of biodiversity shifts.

Biodiversity dependencies: a reliance on or utilizing biodiversity involves using biological resources like materials, liquids, and genetic resources from species, as well as engaging in various ecosystem activities such as pollination, water filtration, and control of crop pests and diseases and water flow regulation.

Table 3.5 Reporting organization requirements for GRI 304: Biodiversity 2016

Categories	Disclosures	Requirements
Topic management disclosures	3-3 Management of material topics in GRI 3 Material Topics 2021	The reporting organization shall report how it manages biodiversity using Disclosure 3-3 in GRI 3: Material Topics 2021 (a) Describe the actual and potential, negative, and positive impacts on the economy, environment, and people, including impacts on their human rights (b) Report whether the organization is involved with the negative impacts through its activities or as a result of its business relationships, and describe the activities or business relationships (c) Describe its policies or commitments regarding the material topic (d) Describe actions taken to manage the topic and related impacts, including: (i) Actions to prevent or mitigate potential negative impacts (ii) Actions to address actual negative impacts, including actions to provide for or cooperate in their remediation (iii) Actions to manage actual and potential positive impacts (e) Report the following information about tracking the effectiveness of the actions taken: (i) Processes used to track the effectiveness of the actions (ii) Goals, targets, and indicators used to evaluate progress (iii) The effectiveness of the actions, including progress toward the goals and targets (iv) Lessons learned and how these have been incorporated into the organization's operational policies and procedures (f) Describe how engagement with stakeholders has informed the actions taken (3-3-d) and how it has informed whether the actions have been effective (3-3-e)

(continued)

Table 3.5 (continued)

Categories	Disclosures	Requirements
Topic disclosures	304-1 Operational sites owned, leased, managed in, or adjacent to, protected areas and areas of high biodiversity value outside protected areas	(a) For each operational site owned, leased, managed in, or adjacent to, protected areas, and areas of high biodiversity value outside protected areas, the following information: (i) Geographic location (ii) Subsurface and underground land that may be owned, leased, or managed by the organization (iii) Position in relation to the protected area (in the area, adjacent to, or containing portions of the protected area) or the high biodiversity value area outside protected areas (iv) Type of operation (office, manufacturing or production, or extractive) (v) Size of operational site in km2 (or another unit, if appropriate) (vi) Biodiversity value characterized by the attribute of the protected area or area of high biodiversity value outside the protected area (terrestrial, freshwater, or maritime ecosystem) (vii) Biodiversity value characterized by listing of protected status (such as IUCN Protected Area Management Categories, Ramsar Convention, national legislation)
	304-2 Significant impacts of activities, products, and services on biodiversity	1. Nature of significant direct and indirect impacts on biodiversity with reference to one or more of the following: (i) Construction or use of manufacturing plants, mines, and transport infrastructure (ii) Pollution (introduction of substances that do not naturally occur in the habitat from point and non-point sources) (iii) Introduction of invasive species, pests, and pathogens (iv) Reduction of species (v) Habitat conversion (vi) Changes in ecological processes outside the natural range of variation (such as salinity or changes in groundwater level) 2. Significant direct and indirect positive and negative impacts with reference to the following: (i) Species affected (ii) Extent of areas impacted (iii) Duration of impacts (iv) Reversibility or irreversibility of the impacts

(continued)

Table 3.5 (continued)

Categories	Disclosures	Requirements
	304-3 Habitats protected or restored	(a) Size and location of all habitat areas protected or restored and whether the success of the restoration measure was or is approved by independent external professionals (b) Whether partnerships exist with third parties to protect or restore habitat areas distinct from where the organization has overseen and implemented restoration or protection measures. (c) Status of each area based on its condition at the close of the reporting period (d) Standards, methodologies, and assumptions used.
	304-4 IUCN Red List species and national conservation list species with habitats in areas affected by operations	(a) Total number of the International Union for Conservation of Nature Red List of Threatened Species (IUCN Red List species and national conservation list species) with habitats in areas affected by the operations of the organization, by level of extinction risk: (i) Critically endangered (ii) Endangered (iii) Vulnerable (iv) Near threatened (v) Least concern

(continued)

The scope of these biodiversity dependencies and impacts varies according to the industry, value chain, and geographic location. Sectors heavily reliant on natural resources, like agriculture, forestry, and fishing, typically yield direct impacts, while tertiary sectors often have indirect interactions through their supply chains. These interplays between dependencies and impacts carry economic implications, resulting in both costs and benefits for businesses and society. These implications include expenses for cleanup, sanctions, the development of plans to mitigate ecological harm, revenue loss due to reputational damage from inadequate biodiversity management, or disruptions in agricultural supply chains stemming from declines in species populations essential to the organization, such as pollinators.

The biodiversity-related disclosure checklist, as summarized and presented in Table 3.6, offers a summary of recommendations for incorporating significant biodiversity-related information into mainstream reports, in accordance with the CDSB requirements. These checklist items are not obligatory mandates but rather suggested disclosures that should be integrated into mainstream reports if they hold material relevance for the organization. For companies at the initial stages of their biodiversity reporting, a phased approach is advisable. This entails reporting on the elements for which the organization currently possesses information, while concurrently outlining a plan to address additional elements in forthcoming periods, aligning with the principles outlined in "Disclosing information in a changing landscape."

Table 3.6 CDSB framework's application guidance for biodiversity-related disclosure checklist

REQ-01 Governance Disclosure shall describe the governance of environmental policies, strategy, and information Does the disclosure: Identify the person(s) or committee responsible for biodiversity-related policies, strategy, and information? Explain how biodiversity-related policies, strategy and management responses are delegated to management? Explain whether there are specific roles or mechanisms in place in priority geographical areas and for priority products/services to tackle compliance with the biodiversity-related regulatory landscape, implementation of biodiversity management responses, and engagement with stakeholders? Describe any systems for accountability and incentivization of biodiversity management? Explain whether the governance mechanisms for biodiversity-related policies, strategies, and disclosure differ from other significant concerns and, if so, why?	*REQ-04 Sources of environmental impact* Quantitative and qualitative results, together with the methodologies used to prepare them, shall be reported to reflect material sources of environmental impact Does the disclosure: Provide a selection of relevant biodiversity impact indicators and metrics, considering sources of material biodiversity impacts, changes to the state of biodiversity, and valuation of impacts? Provide relevant baselines/reference states for metrics and both absolute and normalized metrics where possible? Provide explanations and contextualization of the metrics including the methodologies used, levels of uncertainty, and appropriate narrative to assist understanding of results? Categorize and disaggregate metrics where possible to support understanding and comparability?
REQ-02 Management's environmental policies, strategy, and targets Disclosures shall report management's environmental policies, strategy, and targets, including the indicators, plans and timelines used to assess performance Does the disclosure: Explain the material biodiversity-related dependencies and impacts of the organization with additional consideration of links to natural capital? Summarize the biodiversity policies and strategies and how they support or link to the organization's risks and opportunities and overall strategy? When applicable, explain whether and how biodiversity strategies, policies, and management are established through stakeholder engagement and connect with relevant external societal agreements, policies, and targets? Set out targets (which, where possible, should be contextual, science-based, and time bound), timelines, and indicators for delivery of biodiversity policy and strategy with methods and baselines, including progress towards targets? Detail the resourcing of the delivery and management of water policies and strategies?	*REQ-05 Performance and comparative analysis* Disclosures shall include an analysis of the information disclosed in REQ-4 compared with any performance targets set and with results reported in previous periods Does the disclosure: Provide appropriate historical data to the results reported from REQ-04 for material biodiversity-related impacts to allow for useful comparison, including details on hotspot areas? Contextualize the performance with baselines, targets, and other criteria used to assess progress? Explain the major trends with reference to drivers of change under and/or outside the control of the organization?

(continued)

Table 3.6 (continued)

REQ-03 Risks and opportunities Disclosures shall explain the material current and anticipated environmental risks and opportunities affecting the organization Does the disclosure: Identify material water-related risks and opportunities by adopting a basin-scale and value chain approach and by considering different types of risks? Explain the implications of material water-related risks and opportunities on business and value chains, specifying geographic locations and time horizons in which the risks will materialize? Describe the systems and processes used for assessing, identifying, and monitoring water-related risks and opportunities, including whether they are integrated with existing risk management systems and processes?	*REQ-06 Outlook* Management shall summarize their conclusions about the effect of environmental impacts, risks and opportunities on the organization's future performance and position Does the disclosure: Explain the likely effect of future biodiversity-related impacts, risks, and opportunities as well as of biodiversity strategy on company performance and resilience, taking into account of regulatory and market trends and environmental changes? Identify and explain the time horizons used for reporting on corporate outlook? Explain any techniques, such as scenario analysis, used to inform the outlook including the methods, scenarios and assumptions used, and any shortcomings and uncertainties?

(continued)

3.2 Environmental Impact Assessment by Using LCA

Businesses spanning diverse sectors have established specific objectives centered around environmental, social, and governance (ESG) criteria. To effectively measure progress toward these goals, various approaches have emerged, with greenhouse gas (GHG) emissions taking the forefront as a core ESG metric. This metric frequently serves as a key indicator of environmental impact. Alongside GHG emissions, metrics covering aspects such as embodied energy, recycled content, recyclability, environmental impact in material sourcing, end-of-life materials recovery, and sustainable employment opportunities have gained prominence.

3.2.1 Challenges in ESG Metric Alignment

Despite the widespread adoption of ESG metrics, there is currently a lack of alignment in E score methodologies. This discrepancy was highlighted in a recent review conducted by the Organization for Economic Co-operation and Development (OECD) on ESG. The review focused on scrutinizing E score methodologies to align with the expectations of institutions and investors. Drawing insights from key rating providers, including Thomson Reuters, Morgan Stanley Capital International (MSCI), and Bloomberg, the review investigated the diverse methodologies employed by rating providers in assessing each parameter. This disparity in

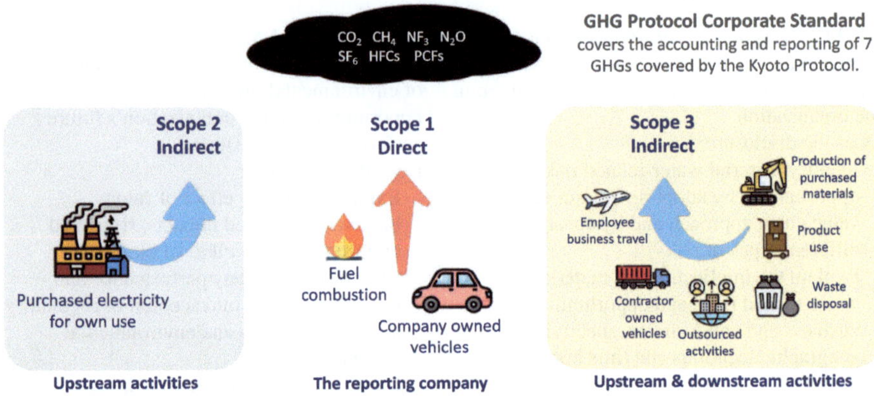

Fig. 3.3 Three scopes of greenhouse gas emissions of a company. (Author's elaboration)

methodologies presents a challenge in achieving standardized and comparable ESG reporting across industries.

The lack of alignment in E score methodologies poses a challenge in achieving standardized and comparable ESG reporting across industries. The Greenhouse Gas Protocol's three source categories (Scope 1, Scope 2, and Scope 3) provide a foundational framework. Scope 1 assesses impacts from sources under the company's ownership or control, Scope 2 from purchased energy sources, and Scope 3 from the extensive supply chain, spanning upstream and downstream components as illustrated in Fig. 3.3. Upholding critical principles such as relevance, completeness, consistency, transparency, and accuracy is essential for high-quality metrics. However, the greatest complexity in measurement lies within the supply chain (Scope 3). The intricate nature of manufacturing supply chains demands more advanced tools like LCA, particularly in achieving rigorous standards in reporting.

3.2.2 LCA: A Foundation for Comprehensive Environmental Assessment

In ESG reporting, a major challenge is ensuring that metrics meet rigorous standards. Recognizing how a company's Scope 3 impacts connect with those of its suppliers or customers opens up an opportunity to enhance reporting standards universally (Engel-Cox et al., 2022). While the current metrics are important, the intricate nature of manufacturing supply chains demands more advanced tools like life cycle assessments (LCA). This provides a unique opportunity for the research community and advanced manufacturers, especially given the increasing demand for robust ESG reporting.

Moreover, companies actively align their environmental practices with sustainability goals by pursuing certifications like LEED. This involves adopting eco-friendly products, permeable paving blocks, single-flush toilets, and solar roofing. The circular economy approach further amplifies the efficient use of materials. Sustainable supply chain strategies and LCA contribute to compliance with environmental regulations such as CBAM and extended producer responsibility (EPR).

Numerous examples underscore the potential for environmentally responsible investments to yield competitive returns. These investments are carefully selected based on a company's operations and alignment with recognized ESG standards, such as those established by the Organization for Economic Co-operation and Development (OECD) and GRI. For instance, a company with substantial energy consumption may set a net-zero transition goal, focusing on reducing energy losses, installing solar panels, implementing smart factory technologies, and securing LEED office building certifications.

Amid the growing trend of ESG investing, achieving sustainable development presents significant challenges, particularly in the environmental (E) aspect. Current environmental initiatives like Renewable Energy 100% (RE100), carbon neutrality, and the circular economy are widely used, but issues such as limited comparability and biased metrics hinder their effectiveness (Senadheera et al., 2021). To make the E pillar more relevant in ESG reporting, a comprehensive set of metrics addressing various environmental aspects is essential. Overcoming these challenges is crucial for making the E pillar a powerful tool for promoting sustainable finance and development.

To ensure the authenticity of these environmental claims and avoid the pitfalls of greenwashing, companies increasingly turn to LCA. LCA offers a comprehensive evaluation of a product or process, considering its environmental impact from raw material extraction to disposal. By utilizing LCA, companies can provide a more accurate and transparent account of their environmental performance improvements, demonstrating a commitment to genuine sustainability. This analytical tool enables a thorough examination of the entire life cycle, guarding against superficial or misleading claims and bolstering the credibility of a company's commitment to ESG goals. Similarly, companies heavily reliant on water resources in their production processes prioritize reducing wastewater generation and responsibly returning clean water to the community, often setting targets like a 15-million-cubic-meter clean water return. Integrating LCA into these sustainability initiatives adds an additional layer of credibility, allowing companies to quantify and communicate the positive impact of their water management strategies across the entire supply chain. This comprehensive approach not only protects against greenwashing but also reinforces the integrity of the company's environmental performance improvements.

Introduction to LCA

LCA stands as a well-established and widely adopted methodology for quantifying the environmental consequences associated with a particular technology. Typically, LCA focuses on the examination of embodied properties that trace their origins upstream through the supply chain.

LCA proves to be a valuable instrument in addressing the environmental dimension of sustainability. According to the International Organization for Standardization (ISO) in the 14040 series, LCA serves as a tool for quantifying the resource consumption and environmental impact contribution throughout the life cycle of a product or service (International Organization for Standardization, 2006). This assessment can be applied to various scales, encompassing products, processes, and even extensive systems. When applied to products, LCA involves quantifying inputs and outputs across the life cycle, converting them into environmental impacts. The calculation of these impacts relies on emission conversion factors, such as those provided by the Intergovernmental Panel on Climate Change (IPCC). The life cycle typically encompasses phases like raw material acquisition, manufacturing, use, and end-of-life for a product or service. LCA is often described as a process from cradle to grave, cradle to gate, or cradle to cradle (McDonough & Braungart, 2002), as illustrated in Fig. 3.4. A cradle-to-grave design considers all life-cycle phases, cradle-to-gate involves raw material acquisition and manufacturing, and cradle-to-cradle includes all phases, with generated waste being recycled and used as a raw material in manufacturing.

To illustrate, consider the production of photovoltaic panels; the first step encompasses the energy required to construct the panel itself, composed of materials like glass, aluminum, and semiconductors. Subsequently, it branches into the energy used to produce the glass and, further downstream, the energy consumed in the extraction and refinement of sand for glass manufacturing, forming a cascading chain of calculations. Often, these calculations encapsulate the embodied energy of a technology, similar to the example presented, or they may encompass embodied CO_2. In some instances, LCAs extend their scope to encompass various other environmental impact assessments, ranging from acidification potential to the impact on abiotic resources, shedding light on how a product might affect critical or sensitive non-living resources.

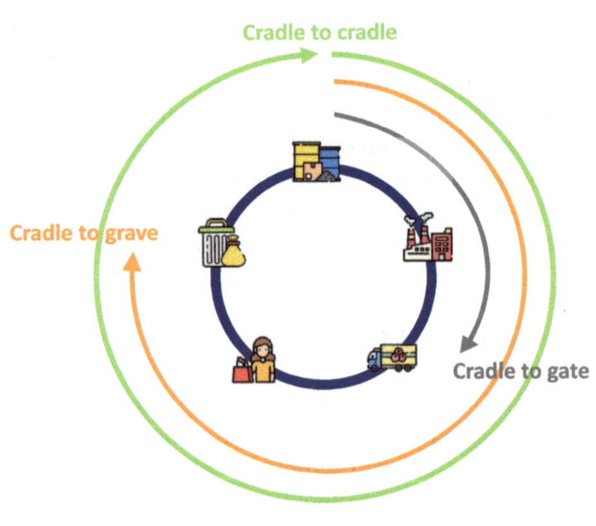

Fig. 3.4 Life cycles of a product or service. (Author's elaboration)

While LCA's strength lies in its adherence to well-accepted guidelines, enhancing the comparability of results through consistent boundary conditions, calculation transparency, and methodology, it has inherent limitations. Notably, it fails to account for the volatility of supply chains, the economic aspects of analysis, and the broader societal impacts associated with the technologies under scrutiny. Furthermore, bolstering transparency in LCA can provide valuable insights, enabling others to gain a comprehensive understanding of the analysis and facilitating its reproducibility. The development of LCA standards, such as those found in ISO 14040-14044, has evolved in tandem with the increasing adoption and interest in LCA evaluations that trace back to the 1960s. However, it remains imperative to acknowledge the areas where LCA may benefit from further refinement and expansion.

LCA Steps and Considerations

ISO's LCA comprises four primary steps: (1) goal and scope definition, (2) life cycle inventory analysis (LCI), (3) life cycle impact assessment (LCIA), and (4) interpretation and improvement of the results. Iterations between these steps are common; for instance, additional inventories may be needed if more products or inputs are included in the system boundary defined during the goal and scope definition step, as depicted in Fig. 3.5. ISO's officially designated reporting mechanisms for LCAs are environmental labels and environmental product declarations (EPDs), outlined by ISO 14025. Product category rules (PCRs) are documents providing regulations, requirements, and guidelines for developing LCAs and EPDs for specific products, ensuring comparability and transparency in LCA studies.

Goal and Scope Definition establishes the study/project's purpose, identifies the audience and stakeholders, specifies the product and its application for examination, determines the level of detail and environmental impact to be considered, sets the system boundaries, and defines the functional unit. The system boundaries specify

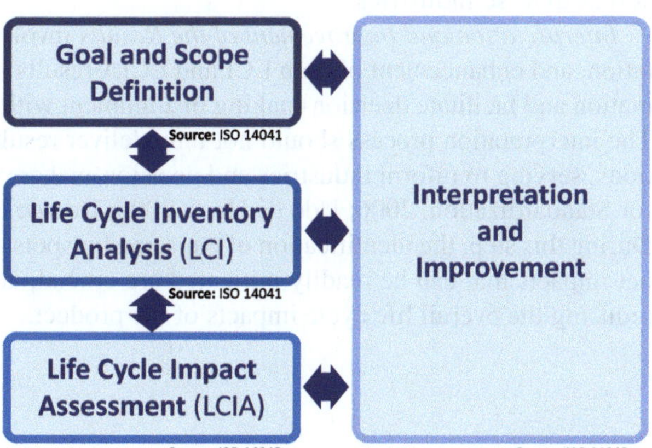

Fig. 3.5 Life cycle assessment steps. (Adapted from International Organization for Standardization, 2006)

the unit process to be incorporated into the modeled system. System boundaries are determined in the goal and scope definition phase, although they may be finalized during the life cycle inventory (LCI) stage as sufficient information becomes available. The functional unit serves as the benchmark for comparison, ensuring it fairly represents the function of the options being compared.

Life Cycle Inventory Analysis constitutes a crucial stage as the quality of LCA results heavily relies on the quality of LCI. LCI can be a time-intensive process, especially when investigating a novel or unconventional process lacking existing data or databases. Inventory data is gathered from diverse sources, either primary (e.g., experiments and questionnaires) or secondary (e.g., literature, government and corporate reports, patents, and life cycle databases such as ecoinvent, ETH-ESU, IdeMat, Franklin, and USLCI (OpenLCA, n.d.; Frischknecht et al., 2007; Frischknecht & Jungbluth, 2004; Norris, 2003; Spriensma, 2004)). These inventories align with the processes specified in the system boundary to attain the defined goals established in the goal and scope definition phase. Multiple life cycle databases are offered by academic institutes, government sectors, and consulting companies. It is crucial to note that transparency and traceability are pivotal aspects in conducting LCIs.

Life Cycle Impact Assessment involves the conversion of LCI data into easily comprehensible and quantifiable environmental impacts, encompassing factors such as global warming, eutrophication, acidification, ozone depletion, and smog formation potentials. This process comprises three main steps: impact category definition, classification, and characterization. Several LCIA tools are utilized for this purpose, including CML 2016, ReCiPe, EPS, IMPACT World +, and TRACI (Tool for Reduction and Assessment of Chemical and other environmental Impacts). For example, global warming potential can be manually calculated using conversion factors for greenhouse gas constituents provided by the IPCC. Beyond this, the LCIA results can be subjected to normalization, grouping, weighting, and analysis, effectively enhancing the quality of the outcomes. To aid in these steps, LCA software like SimaPro or OpenLCA proves invaluable, offering databases containing inventories for materials and processes across diverse industries.

Interpretation and Improvement of the Results involves the correlation, interpretation, and enhancement of both LCI and LCIA results to present meaningful information and facilitate decision-making in alignment with the defined goal and scope. The interpretation process should not only deliver results but also elucidate limitations, serving to inform industries and decision-makers (International Organization for Standardization, 2006; Udo de Haes, 2006; Udo de Haes & van Rooijen, 2005). During this step, the identification of emission hotspots and the recognition of product impacts that can be readily mitigated are crucial, aiming to offer strategies for reducing the overall life cycle impacts of the product.

3.2.3 Application of LCA in Environmental Impact Assessment

Life cycle approaches involve the consideration of a product or service throughout its entire life cycle, typically comprising four phases: raw material acquisition, use, manufacturing, and end-of-life. Widely employed for quantifying life-cycle environmental impacts and costs, life cycle assessment or analysis (LCA) provides three key benefits: (1) enhancing product or service performance, (2) preventing unintended consequences, and (3) facilitating the decision-making process (Soratana et al., 2021).

1. To enhance product or service performance, the life cycle approach identifies hotspots—processes with significant costs, resource consumption, or environmental impact. For example, in the hotel industry, energy consumption sources like air conditioners, lighting, televisions, and water heaters are monitored to pinpoint hotspots. This allows for improvements, such as installing thermal insulation materials or energy-efficient air conditioners, thereby reducing costs and environmental impacts (Soratana et al., 2021).
2. Considering a product or service from a life cycle perspective helps avoid unintended consequences. An illustration is the shift from fossil fuel to first-generation biofuels, which mitigates global warming potential but leads to eutrophication during sugarcane and corn cultivation. Similarly, switching from single-use plastic to biomaterial containers may reduce solid waste but increase greenhouse gas emissions due to bio-material waste mismanagement. Life cycle approaches are instrumental in mitigating issues like overtourism caused by unregulated policies, enabling a thorough examination of potential adverse impacts (Soratana et al., 2013).
3. In the decision-making process, life cycle approaches contribute by providing quantitative data for comparison. Constructing comprehensive inventories allows for quantitative assessments, although trade-offs exist. While life cycle approaches should be coupled with weighting techniques for specific tourism destinations, their adoption is crucial for managing sustainability issues across all three pillars—economic, environmental, and social—in the industrial sector.

3.3 Incorporating LCA into ESG Reporting

Incorporating LCA into ESG reporting emerges as an important strategy, empowering companies committed to comprehensive sustainability disclosure. This systematic framework allows organizations to evaluate the environmental footprint of products, processes, or services throughout their life cycle, deepening their understanding of environmental performance and enhancing transparency with stakeholders. The integration of LCA not only bridges the gap between sustainability goals and credible reporting but also improves environmental metrics, enabling the precise quantification of carbon emissions, water usage, and waste generation for key performance indicators (KPIs) and ESG reporting targets.

Recognizing the need to address gaps in ESG reporting, a shift from corporate-level to a product-level approach becomes increasingly significant with the incorporation of LCA. This adaptation enables companies to make substantial progress toward decarbonization targets, a crucial step in achieving net-zero goals. Harmonizing with ESG reporting frameworks such as the Global Reporting Initiative (GRI) and the Sustainability Accounting Standards Board (SASB), the integration of LCA into ESG reporting ensures compliance and relevance in the evolving ESG disclosure landscape. Renowned for its structured methodology, LCA offers a systematic approach to measure, report, and verify environmental data, enhancing companies' alignment with sustainability goals and fostering a comprehensive understanding of their environmental impact.

Moreover, the combination of LCA with the product life cycle enhances the evaluation of ESG growth, distinguishing between ESG growth and financially ESG sustainable growth (Bellandi, 2022). This holistic approach links specific LCA cycles to the business's product life cycle, allowing for the allocation of ESG weights to various life cycle phases and enabling a thorough ESG rating assessment at each stage. The integration of LCA extends beyond the traditional engineering perspective, encompassing a broader "business" concept of life cycle used in management and business studies.

ESG and LCA, when synergized, form a vital consolidation for unlocking the potential of science-based environmental savings and meeting the escalating demand for transparent information regarding the environmental impact of products and services. A comprehensive evaluation of a product's impact throughout its life cycle, exemplified by the analysis of reusable packaging containers (RPCs) of IFCO, the world's leading supplier of RPCs, underscores the environmental savings and advantages of RPCs over single-use packaging. The incorporation of ESG and LCA into a circular model for pooling RPCs aligns with sustainability goals, offering an end-to-end, durable, and reusable packaging solution. This circular approach, revolving around the IFCO SmartCycle™—a distinctive closed-loop "share-and-reuse" RPC pooling system—facilitates emissions reduction in the global fresh grocery supply chain. An ESG and LCA lens applied holistically to the fresh grocery supply chain reveals areas for potential science-based environmental savings.

The commitment to transparency and independence in LCAs is exemplified by entrusting the in-depth analysis of IFCO reusable packaging containers to Franklin Associates. Aligning with international standards, particularly ISO 14044 compliance and peer review by experts, ensures the credibility of the LCAs conducted by Franklin Associates. This commitment underscores the dedication to sustainability, reinforcing the positive influence made on the industry.

3.3.1 Benefits and Challenges

Benefits The integration of LCA into ESG reporting offers numerous advantages.

- *LCA integration offers advantages certified by ISO 14040, aligning with the evolving trend of ESG toward science-based target.*

LCA is a certified methodology according to ISO 14040. Recognized by the International Organization for Standardization (ISO), LCA, as outlined in ISO 14040: 2007, stands as a dependable tool for environmental impact assessment. This addresses a significant gap in the E pillar of ESG, where standardized methods are currently lacking across different ratings such as MSCI, S&P, and Bloomberg (Senadheera et al., 2021). Furthermore, as a quantitative method, LCA provides results in numerical form, aligning seamlessly with the evolving trend of ESG toward science-based targets.

- *Companies gain the ability to identify environmental hotspots and focus on specific areas for environmental improvement.*

By utilizing LCA, companies not only gain the ability to identify environmental hotspots or processes with substantial impact contributions but also facilitate goal achievement by directing focus on specific areas for environmental improvement. This comprehensive insight, provided by LCA, extends beyond the assessment of carbon emissions, quantifying various environmental impacts such as eutrophication and photochemical smog formation. This addresses the evolving importance of a thorough environmental evaluation.

In response to these challenges, companies are increasingly leveraging LCAs and technology to strategically enhance their ESG performance, with a particular emphasis on supply chain optimization. LCAs contribute significantly to performance improvement by aiding in product selection for market entry, evaluating value from supplier engagement, optimizing operational efficiency, and enhancing marketing endeavors. What sets LCAs apart is their proactive utility, serving not only for retrospective analysis but also as a tool for shaping a sustainable future. The proactive nature of LCAs aligns with the growing emphasis on science-based targets and contributes to a holistic approach to environmental and sustainability goals.

- *LCAs contribute significantly to performance improvement in supply chain optimization, product selection, and marketing endeavors.*

The integration of LCA further facilitates data comparison across companies, ensuring consistency and alignment in LCA calculations and documentation. The distinctive aspect of LCAs is their proactive utility, not confined to retrospective analysis but also serving as a tool for shaping a more sustainable future, enabling informed decisions for the improvement of the environment and society. Additionally, LCA reinforces a company's decision-making process when exploring environmental opportunities, such as transitioning to renewable energy sources, adopting resource-conserving processes, and minimizing pollution to reduce carbon footprint. Considering the E pillar not only aligns with regulatory expectations but also creates competitive advantages for eco-friendly products and services.

- *Facilitates data comparison across companies, ensuring consistency and alignment in LCA calculations and documentation.*

Despite the potential capital costs associated with renewable and alternative technologies, especially challenging for smaller companies, a transparent commitment to environmental sustainability enhances the prospects of attracting investors focused on sustainability and tapping into green business opportunities. In essence, integrating LCA into ESG reporting not only meets standards and regulatory requirements but also positions companies strategically for a sustainable and competitive future.

Challenges While LCA holds the potential to significantly contribute to ESG reporting, it is not without its challenges.

- *Conducting comprehensive LCAs is time-consuming and demands extensive collaboration across the supply chain.*

The process of conducting a comprehensive LCA is inherently time-consuming and demands extensive collaboration and effort from stakeholders across the supply chain. This collaboration is essential for obtaining the diverse data needed to assess the environmental impact thoroughly.

- *Lack of data, especially in emerging technologies, poses a hurdle, but opportunities exist to leverage technologies like AI and blockchain for streamlined data collection.*

The lack of data and information, particularly in the context of emerging technologies, poses a significant hurdle for companies undertaking LCA. Despite these challenges, there exists an opportunity for businesses to address these issues by leveraging technologies such as AI and blockchain to develop databases that streamline the data collection and computation processes for LCA.

During the infancy stage, when product knowledge is preliminary and processes are introductory, the LCA steps may differ compared to the maturity phase. In the latter phase, companies may encourage varied product usage, modifications, or different applications, prompting shifts in customer targets, distribution channels, and production processes. Technological advancements and maturation can introduce new techniques, influencing the lifespan and end-of-life recovery value of a product. Quality, design flaws, and the need for service and remanufacturing also evolve across life cycle stages, potentially posing challenges in servicing second-hand products during the decline phase due to a lack of spare parts (Bellandi, 2022; Cao & Folan, 2012). These complexities underscore the intricate nature of conducting LCAs throughout the varying phases of a product's life cycle.

- *Variability in processes, methods, and durations across different life cycle stages underscores the complexity of conducting LCAs.*

To understand the complexities involved, consider a product undergoing a standard LCA, encompassing stages from raw materials to recycling. Across the four phases of the life cycle model, different steps may be performed based on production and sales fluctuations. The critical factor lies in the variability of processes, methods, durations, and associated ESG loads chosen at each stage. This variability arises due to shifts in strategic focus dictated by the product's configuration, features, and the business model employed at each phase. This dynamic nature of processes and methods across different life cycle stages underscores the inherent complexity involved in conducting LCAs.

3.4 Summary

Environmental performance measurement is essential for driving sustainability across industries, with LCA serving as a crucial tool to evaluate the environmental impacts of products and services throughout their entire life cycle. LCA involves four primary steps, goal and scope definition, life cycle inventory analysis, life cycle impact assessment, and interpretation and improvement of the results, each providing valuable insights into a product's environmental footprint. Despite its transparency and ability to enable comparisons, LCA does have limitations, such as its inability to account for supply chain volatility and broader societal impacts.

Integrating LCA into ESG reporting allows organizations to effectively measure and disclose their environmental performance, aligning with globally recognized frameworks like the Global Reporting Initiative (GRI) and the Sustainability Accounting Standards Board (SASB). This integration supports sustainability goals, facilitates data-driven decision-making, and enables organizations to make informed choices. However, challenges such as data gaps and the need for cross-sector collaboration remain, though technologies like AI and blockchain offer promising solutions to streamline the data collection and analysis processes.

In this context, challenges in environmental sustainability methodologies, especially related to GHG emissions categorization (Scope 1, 2, and 3) and supply chain measurement, are prominent. As highlighted by the OECD, life cycle assessments (LCA) continue to be a vital tool, helping companies identify environmental hotspots and navigate data-related challenges despite their time-consuming nature. Integrating LCA into ESG reporting not only addresses these challenges but also strategically positions companies for a more competitive and sustainable future.

3.5 Discussion Questions

1. How can life cycle assessment (LCA) be effectively integrated into supply chain management to identify and address environmental hotspots? What role does collaboration across industries or sectors play in this process?

2. What are the key limitations of LCA in capturing supply chain volatility and broader societal impacts? How can emerging technologies like AI and blockchain help overcome these challenges?
3. Incorporating LCA into ESG reporting frameworks like GRI and SASB is said to enhance transparency and credibility. What specific advantages does this integration provide, and how might it benefit organizations in competitive markets?
4. Organizations often face challenges in obtaining comprehensive environmental data due to data gaps, especially for new or emerging products. What strategies can help balance the need for detailed data with these limitations?
5. Provide practical examples of how LCA has driven sustainability improvements in sectors like manufacturing, hospitality, or agriculture. What lessons can other industries draw from these examples to adopt LCA more effectively?
6. Challenges in aligning environmental scoring methodologies, particularly for metrics like GHG emissions (Scope 1, 2, and 3), impact the credibility of ESG criteria. How can companies and regulators work together to improve consistency and comparability in reporting?

References

Bellandi, F. (2022). Further developments of the financially-ESG sustainable growth matrix. *International Journal of Business and Management, 17*(2). https://doi.org/10.5539/ijbm.v17n2p112

Cao, H., & Folan, P. (2012). Product life cycle: The evolution of a paradigm and literature review from 1950–2009. *Production Planning & Control, 23*(8), 641–662. https://doi.org/10.1080/09537287.2011.577460

Climate Disclosure Standards Board (CDSB). (2021a). *CDSB Framework: Application guidance for biodiversity-related disclosures.*

Climate Disclosure Standards Board (CDSB). (2021b). *CDSB Framework: Application guidance for water-related disclosures.*

OpenLCA. (n.d.). IdeMat online database. Retrieved March 5, 2025, from https://nexus.openlca.org/database/IDEMAT

Engel-Cox, J. A., Wikoff, H. M., & Reese, S. B. (2022). Techno-economic, environmental, and social measurement of clean energy technology supply chains. *Journal of Advanced Manufacturing and Processing, 4*(3), e10131. https://doi.org/10.1002/amp2.10131

Frischknecht, R., & Jungbluth, N. (2004). *SimaPro database manual: The ETH-ESU 96 libraries* (Version 2.1). PRé Consultants

Frischknecht, R., et al. (2007). *Code of practice: Ecoinvent report No. 2 (Data v2.0).* Swiss Centre for Life Cycle Inventories

International Organization for Standardization. (2006). *ISO 14040:2006, environmental management - life cycle assessment - principles and framework.* International Organization for Standardization.

McDonough, W., & Braungart, M. (2002). *Cradle to cradle: Remaking the way we make things.* North Point Press.

Norris, G. A. (2003). *SimaPro database manual: The Franklin US LCI Library* (Version 2.0). PRé Consultants

Senadheera, S. S., Withana, P. A., Dissanayake, P. D., Sarkar, B., Chopra, S. S., Rhee, J. H., & Ok, Y. S. (2021). Scoring environment pillar in environmental, social, and governance (ESG)

assessment. *Sustainable Environment, 7*(1), 1960097. https://doi.org/10.1080/27658511.2021.1960097

Soratana, K., Khanna, V., & Landis, A. E. (2013). Re-envisioning the renewable fuel standard to minimize unintended consequences: A comparison of microalgal diesel with other biodiesels. *Applied Energy, 112*, 194–204. https://doi.org/10.1016/j.apenergy.2013.05.082

Soratana, K., Landis, A. E., Jing, F., & Suto, H. (2021). *Supply chain management of tourism towards sustainability* (1st ed.). Springer.

Spriensma, R. (2004). *SimaPro database manual. PRé Consultants*. https://simapro.com/wp-content/uploads/2020/10/DatabaseManualMethods.pdf?utm_source=chatgpt.com

The Global Sustainability Standards Board (GSSB). (2022). *GRI 304: Biodiversity 2016*.

The Global Sustainability Standards Board (GSSB). (2023). *GRI 303: Water and effluents 2018*.

Udo de Haes, H. (2006). How to approach land use in LCIA or, how to avoid the Cinderella effect? *The International Journal of Life Cycle Assessment, 11*(4), 219–221. https://doi.org/10.1065/lca2006.07.257

Udo de Haes, H., & van Rooijen, M. (2005). *Life cycle approaches: The road from analysis to practice*. Retrieved from Nairobi/Pensacola, https://www.lifecycleinitiative.org/wp-content/uploads/2012/12/2005%20-%20LCA.pdf

[illegible]

Chapter 4
Social and Corporate Governance Factors in ESG Investing

Nataliia Tkalenko

Abstract This chapter delves into the intricate dynamics of corporate governance within the scope of environmental, social, and governance (ESG) investing, underscoring its pivotal role in shaping investment decisions. It lays out the foundational principles of corporate governance, including shareholder rights, equitable treatment of shareholders, stakeholder roles, disclosure, transparency, and the board of directors' responsibilities. The chapter pays special attention to the implementation of sustainable development principles within corporate activities and highlights the importance of integrating ESG factors into these governance frameworks.

In developed countries, the system of relationships between key corporate players—such as shareholders, managers, directors, creditors, employees, and others—is well-established, aiming to maximize corporate efficiency, attract investments, and fulfill legal and social obligations. The chapter differentiates between corporate management and governance, with the latter encompassing a broader interaction among various entities impacting a firm's functioning. It emphasizes that corporate governance is not just about business operations but also involves creating long-term value through aligning financial and social performance and ensuring accountability and legitimacy in the eyes of society.

The chapter introduces a four-pronged model of corporate governance, encompassing people, purpose, process, and performance, each playing a vital role in the organizational structure. It also discusses the importance of shareholder rights, equal treatment of all shareholders, the role of stakeholders in governance, transparency in corporate operations, and the critical responsibilities of the board of directors in steering the company's strategic direction.

Finally, the chapter addresses the challenge of ensuring that the interests of shareholders, who are the corporation's owners, are observed in an environment where significant decision-making information is often asymmetrically distributed in favor of managers. It explores the internal dynamics of corporations and their

N. Tkalenko (✉)
Department of Management, Mendel University in Brno, Brno, Czech Republic
e-mail: nataliia.tkalenko@mendelu.cz

M. Bednárová, K. Soratana (eds.), *Environmental, Social, and Governance (ESG) Investment and Reporting*, https://doi.org/10.1007/978-3-031-84235-1_4

interactions with external environments as potential capital sources, highlighting the varying forms of corporate governance across different countries.

4.1 Evaluating Social Responsibility and Its Impact on Investment Decisions

Beginning in the second half of the twentieth century, despite the existence of normative legal acts and a series of initiatives, significant progress in combating climate change has not been achieved. For a long time, the approach to climate change and social issues as voluntary corporate social responsibility did not find the necessary support in the business sector. In essence, the concept itself was grounded in 1953 by scientist Howard Bowen in his work *Social Responsibilities of the Businessman* (Bowen, 1953) and envisaged the businessman's obligation to follow such a policy, make decisions, or follow a sequence of actions that are desirable in terms of the goals and values of society. Meanwhile, the initial ideas regarding the necessity of such responsible behavior can be found even earlier in George Perkins' work *The Modern Corporation* (1908). In this work, the author concluded that the larger a corporation becomes, the greater its obligations to society are.

The 1950s were a turning point in the field of workers' rights protection, as it was during this time that the activities of public and international organizations, as well as civil society, were intensified with the aim of creating equal conditions for work and the opportunity to freely and actively participate in union activities.

However, it was only in the 1980s that the concept of corporate social responsibility (CSR) gained widespread use among companies and other economic agents—companies began to pay more attention to social and environmental issues and spend money on charity, which contributed to improving their image in society. At the same time, all these actions were separate from business activities and did not provide the opportunity to attract external funding for related projects (Carroll, 1991). Additionally, it was difficult for companies to feel quantitative improvements for financial reporting (or at least the connection was indirect and hard to evaluate).

Considering these shortcomings, in 2011 economists M. Porter and M. Kramer proposed a new version of corporate social responsibility—CSV ("creating shared value"). This concept tried to demonstrate the contribution and how a company can improve social and economic conditions within a community or country. According to its authors, it was an attempt to move from agreements to real actions that allow creating shared value as a result of relevant efforts (actions).

It was only in the late 1990s, following the signing of the Kyoto Protocol to the United Nations Framework Convention on Climate Change, that economist J. Elkington managed to develop a mechanism that allows combining the interests of all financial participants and ensuring the inflow of capital to projects that should contribute to the reduction of greenhouse gas emissions and, consequently, improve social development indicators not only for enterprises but also for the state.

In his book *Cannibals with Forks: The Triple Bottom Line of twenty-first Century Business*, the author, based on the analysis of shortcomings of previous concepts, demonstrated the possibility of using existing environmental and social risks to develop new business models (emergence of so-called environmental and social entrepreneurs) and proposed using ESG (environmental, social, and governance) principles to assess investment projects and company activities.

The advantages of this concept are, firstly, the possibility to move from a voluntary approach in combating global threats to new business models, investment, and lending directions in the financial market, secondly, to ensure the formation of a real asset for companies as a result of implementing projects in the aforementioned areas, which can be an additional incentive for investors and creditors to participate in such projects (Cambridge Institute for Sustainability Leadership, 2024).

However, for companies and other economic agents, the use of the provisions of this concept is associated with certain challenges:

- Additional costs (especially, transaction costs);
- New approaches to accounting and reporting of nonfinancial results (reporting on nonfinancial results)
- A more complex system of monitoring and evaluating the obtained results

Corporate social responsibility (CSR) refers to the voluntary contribution to the development of society in social, economic, and environmental spheres, often not directly related to the core activities of companies and exceeding the legal minimum and societal ethical norms. It involves responsibilities toward business partners, employees, local communities, and the population. Unlike legal responsibility, social responsibility entails a certain level of voluntary response from companies to societal issues that go beyond legally defined requirements or ethical norms accepted in society.

According to the European Commission, corporate social responsibility is a concept reflecting the voluntary commitment of companies to participate in improving society's well-being and protecting the environment. Many specialists equate the terms "corporate social responsibility" and "social responsibility." H. Gordon Fitch, a specialist in corporate social responsibility, noted in the 1970s that social responsibility is an attempt to address social problems wholly or partially caused by the activities of companies. Therefore, from our perspective, the concept of "corporate social responsibility" is broader as it encompasses not only the social aspect but also the economic and environmental spheres.

An important characteristic of corporate social responsibility is its multilevel structure. Three main levels of social responsibility can be distinguished: Basic implies a minimum set of characteristics of socially responsible business behavior, such as producing quality products, timely payment of wages, full and timely tax payments, compliance with legislation, environmentally friendly practices, adherence to occupational health and safety requirements, and more. Expanded involves developing partnership relations within entrepreneurial structures based on negotiation processes and is directly related to considering diverse socioeconomic interests of the entire company's collective. In addition to the basic obligations of socially responsible business behavior at the second level, it includes providing employees with a complete social package and additional social guarantees. Comprehensive includes the participation of entrepreneurial structures in financing social projects and programs that go beyond the company's core activities. The object of social responsibility at this level includes not only employees but also individuals not directly associated with the company's activities but having a certain relationship to the territory of its presence. This encompasses socially responsible actions aiming to create and develop favorable socioeconomic conditions for its operation and the participation of entrepreneurial structures in municipal, regional, national, and international projects and programs.

The expansion of business social responsibility and the increased impact on corporate reputation necessitate the assessment of this influence. To investigate how a company's responsible business practices affect its market value and key financial performance indicators, several assumptions are put forward.

Firstly, socially responsible investing influences a company's value expressed by the economic value added (EVA) metric. Currently, the concept of value-based management is relevant in the business environment, with the goal of maximizing the company's value. It is assumed that responsible business conduct will have a positive impact on the company's value, as socially responsible investing is a global trend highly significant to investors. Companies with high ESG ratings are likely to have a higher market value. Economic value added (EVA) is chosen as a criterion for evaluating the market value of the company, serving as a measure of its overall worth. EVA reflects not only positive financial results but also the profitability of invested capital, surpassing the costs of its acquisition and creating additional value.

Secondly, socially responsible investing affects the company's value expressed by market capitalization. Market capitalization, generated using Thomson Reuters data, is calculated based on an algorithm using publicly available stock market data. This indicator is used to measure the company's value since it is formed based on real market data on the company's stock value, unlike EVA, which is predominantly calculated based on balance sheet indicators.

Thirdly, the level of socially responsible business conduct influences the EBITDA, weighted average cost of capital (WACC), and return on assets (ROA) of the company. EBITDA is a commonly used performance indicator for managers and key executives and is also used in calculations of other financial ratios. The impact of corporate social responsibility on financial indicators has been analyzed by various scholars over an extended period. Social investments by a company will positively affect its financial performance essentially indirectly—through the improvement of the company's reputation/image among stakeholders, increased certainty regarding cash flows, and, in some cases, through cost reduction. The profitability of assets can be traced to the relationship between business social responsibility and the efficient use of the company's assets. A socially responsible company changes its goals and strategy and aims to optimize resources for their achievement. It is assumed that the larger the scale of the business, the more opportunities the company must invest in socially significant activities, and the greater the responsibility and commitment of management to conduct business transparently.

There is no single criterion for evaluating the degree of corporate social responsibility (CSR). Various major analytical agencies, companies, investment funds, and stock exchanges create their own rankings of socially responsible companies. Typically, these rankings are based on data from companies published nonfinancial reports. It's worth noting that, unlike International Financial Reporting Standards (IFRS) for financial reporting, there are no universally accepted, strict standards for the presentation of nonfinancial reports. There are only developed recommendations, and the content of such nonfinancial reports essentially relies on the company's discretion. Additionally, internal documents such as corporate governance and ethics codes, anti-corruption policies, quality control, environmental and occupational safety policies, risk assessment, and management approaches, among others, are used as sources of information for compiling CSR rankings. All available information from the media is also utilized.

Corporate social responsibility models in global practice are divided into open and closed (invisible) models. Open models involve corporations taking responsibility for solving issues that are in the interest of society. The open model of CSR is clearly presented in the company's development strategy. Closed (invisible) CSR models reflect the presence of official and unofficial institutions in a country.

The European model of public interest aligns with companies and is enforced by its mandatory adherence. This model signifies values, norms, and rules that companies must adhere to during the implementation of CSR. Corporate social responsibility is primarily a system of mechanisms of state regulation. The British model of CSR combines elements of the US model and continental Europe. The state and public institutions are involved in the process of coordinating public interests, as well as promoting and disseminating best practices.

Corporate social responsibility in the United States began to actively develop in the nineteenth century. Examples include the establishment of public libraries by Rockefeller and public initiatives by Carnegie. Most measures implemented by American corporations are voluntary. In the United States, there are numerous institutions addressing social issues, such as corporate foundations addressing social problems through corporate contributions. The increase in the level of CSR by American corporations is incentivized by legislation that provides corresponding tax incentives. An example is the well-known corporate foundation of Bill Gates, which actively works to improve education and healthcare systems globally.

In Europe, economic responsibility is more focused on business profitability and corporate responsibility to its shareholders. Special attention is given to companies' responsibility to their employees.

In Europe, philanthropy among companies is not as popular and widespread as in the United States. In the United Kingdom, companies began to pay special attention to the development of corporate social responsibility after the reforms of Margaret Thatcher. Currently, a consulting market in the field of corporate social responsibility has emerged. The number of socially responsible investment companies is growing year by year. *The Times* newspaper publishes a Corporate Social Responsibility Index in the Company Profile section every week. Research shows that every year the volume of social investments and the quality of the organization of the social investment process in the corporate sector worldwide are increasing. Being a "corporate citizen" means adhering to all legal obligations to the government, timely and transparent payment of taxes, contributing to the state budget, working honestly and transparently with all stakeholders (clients, business partners, shareholders, investors, suppliers, government entities). It also means adhering to modern standards of corporate governance, openness, and transparency, creating decent working conditions for employees, considering salary levels, implementing modern work organization methods, and developing a corporate culture. It involves maintaining high safety standards, providing equal opportunities for employees, fairly evaluating everyone's contribution to the success of the business, investing in human resources, providing training opportunities for employees, and working toward reducing the negative impact of production cycles on the environment (land, air, water, and ecosystems). Additionally, it involves investing in environmentally friendly production technologies, improving the environmental characteristics of the business, and striving for more rational use of natural resources.

4.2 Corporate Governance and Its Importance in Investment Decisions

In developed countries, the foundations of the system of relations between the main actors of corporations such as shareholders, managers, directors, creditors, employees, suppliers, buyers, etc. are now clearly defined. Such a system is created to solve three main tasks of the corporation: ensuring its maximum efficiency, attracting investments, and fulfilling legal and social obligations. From this point of view, it is impossible to equate the concepts of corporate management and corporate governance.

The first term refers to the activity of professional specialists in the course of conducting business operations. In other words, management is focused on the mechanisms of conducting business. The second concept is much broader: it means the interaction of many individuals and organizations related to the most varied aspects of the firm's functioning. Corporate governance is at a higher level of company management than management. The intersection of the functions of corporate governance and management takes place only in the development of the company's development strategy.

Corporate governance lies at the heart of creating long-term value through the alignment and enhancement of financial and social performance, as well as by ensuring accountability to stakeholders and creating legitimacy for the business in the eyes of society. Corporate governance includes four principles (Fig. 4.1).

- People: These are the organizers who determine the goal to strive for, develop a consistent process for achieving it, evaluate the results of their work, and use these results for development (founders, board, stakeholders, and consumers).

Fig. 4.1 4P model of corporate governance. (Author's elaboration, based on Samans and Nelson (2020))

- Purpose: These are the guiding principles of the organization, a statement of its mission. Each of the company's policies and projects should exist to promote this purpose.
- Process (management process): This is the process by which people achieve the goals of their company. It evolves through performance analysis. Processes are improved over time to consistently achieve their purpose.
- Performance: Performance analysis is a key skill in any industry. The ability to look at the results of the process and determine whether it was successful (or sufficiently successful) and then apply these findings to other areas of activity is one of the main functions of the management process.

In April 1999, a special document approved by the Organisation for Economic Co-operation and Development (which unites 29 countries with developed market economies) formulated the following definition of corporate governance: "Corporate governance refers to the internal means of conducting and controlling corporate activities... One of the key elements for increasing economic efficiency is corporate governance, which includes a complex of relationships between the management (administration) of the company, its board of directors (supervisory board), shareholders, and other interested parties (stakeholders)" (Samans & Nelson, 2020). Corporate governance also determines the mechanisms by which the company's objectives are formulated, the means of achieving and controlling its activities. The document also detailed the five main principles of proper corporate governance (Fig. 4.2).

Shareholder rights (the corporate governance system should protect the rights of stock owners). Protecting shareholder rights is a fundamental aspect of corporate governance. This includes respecting the rights of minority shareholders, ensuring

Fig. 4.2 Main principles of proper corporate governance. (Author's elaboration, based on Samans and Nelson (2020))

equal treatment for all shareholders, and providing opportunities for shareholders to effectively participate in key corporate decisions, such as the election of board members. Active interaction with shareholders can provide valuable information and help win investor trust.

Equal treatment of shareholders (the corporate governance system should ensure equal treatment for all stock owners, including small and foreign shareholders).

The role of stakeholders in corporate governance (the corporate governance system should recognize the legally established rights of stakeholders and encourage active cooperation between the company and all stakeholders to increase public wealth, create new jobs, and achieve financial stability in the corporate sector). The ethical tone of the company, often set by its leadership, plays an important role in corporate governance. A culture of honesty and ethical behavior should permeate all levels of the company. Policies on issues such as conflict of interest, corruption, and corporate social responsibility reflect the company's commitment to ethical behavior.

Disclosure of information and transparency (the corporate governance system should ensure timely disclosure of reliable information about all significant aspects of corporate functioning, including information about financial position, performance results, ownership composition, and management structure). Transparency in corporate governance implies timely and accurate disclosure of all significant issues concerning the corporation, including its financial position, performance, ownership, and management (Harvard Law School Forum on Corporate Governance, 2021). Accountability requires that individuals or groups in the company (such as management or the board of directors) be responsible for their actions and decisions and must explain them to those concerned.

Responsibilities of the board of directors: the board of directors provides strategic guidance for the business and effective control over the work of managers and is accountable to shareholders and the company as a whole. The board of directors plays a central role in corporate governance. The composition, diversity, independence, and experience of its members are important for effective supervision. A diverse and independent board of directors is more likely to challenge managerial decisions and offer different viewpoints, leading to more effective decision-making. The role of the board of directors includes determining the strategic direction of the company, controlling management, and ensuring the company's accountability to shareholders and other stakeholders.

4.3 Evaluating the Role of Corporate Governance in Fostering Sustainable Practices

The problem of corporate governance boils down to creating mechanisms that would ensure the observance of the interests of shareholders, who are the owners of the corporation, in conditions where significant decision-making information (both current and strategic) is asymmetrically distributed in favor of managers often pursuing their interests. Corporate governance includes the consideration of two issues: internal life of the corporation (creation, liquidation, shareholder rights, management body competencies); the corporation's interaction with the external environment, serving as a potential source of capital (issuance of shares, bonds; conditions for acquiring large share packages).

If we consider country-specific features, the forms of corporate governance organization also vary in different countries. The main difference lies in how the distribution of functions between the board of directors and executive bodies is structured. Moreover, the degree of involvement of various stakeholders in the management process also differs. The main participants in corporate relations are:

- Shareholders—are investors in the organization, interested in receiving dividends and a high share price in the event of their sale
- Hired managers—carry out all the main management functions
- Organization staff—directly participate in the production and economic activities of the organization
- Government authorities—form the normative legal basis for corporate relations
- Creditors—participate in financing, production, economic, and other activities of the corporation
- Regional authorities and local communities

From a business perspective, formalized regulation of corporate relations should be envisaged. The final form of the system is shaped by the general economic situation in the country, but the peculiarities of the national culture and the level of stock market development are equally important. Together, all these factors allow us to identify three main models of modern corporate governance (Fig. 4.3).

In today's globalized world, transnational corporations are pivotal in creating and amassing substantial investment capital streams. This elevates corporate governance issues to a significant level, encompassing both national and international domains. The 1990s marked a shift in focus toward safeguarding investor rights and interests. This movement has been supported by various governments and international entities, including the World Bank, the European Bank for Reconstruction and Development, the International Finance Corporation, and the Organisation for Economic Co-operation and Development. These bodies have been instrumental in formulating and implementing various national codes for corporate governance.

Influenced by challenges faced by large corporations in the United States, the United Kingdom, and Canada, the development of these corporate governance codes was initiated. These efforts saw collaboration from stock exchange representatives, corporate entities, institutional investors, and associations of directors and corporate managers, alongside groups dedicated to defending investor rights. These initial codes served as a template for other nations to create their own corporate governance frameworks. By the end of 2002, approximately 90 different corporate governance codes had been established across various countries and corporate entities.

Corporate governance that meets global standards allows companies to enter the world IPO markets and participate in international capital markets. This process is associated with an increase in the value of the enterprise, which is expressed in capitalization in the presence of a liquid stock market, or in liquid price in its absence.

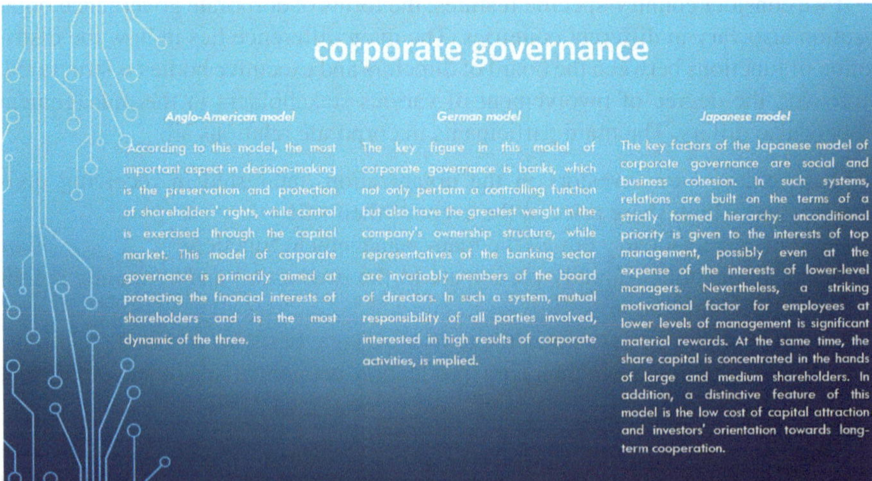

Fig. 4.3 Models of modern corporate governance. (Adapted from Samans and Nelson (2020))

In modern conditions, increasing the value of the business is the main goal of enterprise owners and its management in the long term.

The main principles of corporate governance are closely related to the components of maximizing enterprise value. This is primarily manifested in investment attractiveness, as the main aspect of corporate governance practice is associated with ensuring the inflow of external capital. The principle of information disclosure and transparency involves providing regular and comparable information that allows for effective control over the activities of the enterprise and the assessment of the quality of operational management. This principle, through the methods of value management, allows for the evaluation of the corporation's management and provides a basis for the valuation of securities. Owners, through the Board of Directors, exercise continuous control over the management activities of the management, and value management methods allow identifying key factors of value and mechanisms for its creation or destruction.

Analyzing the indicators of value creation allows the Board of Directors and the general meeting of shareholders to control the effectiveness of management. Important strategic decisions are analyzed in terms of their impact on the value assessment of the business, which allows for an integrated indicator on various time horizons and considers the impact on different business areas. For management, these decisions form the basis for the distribution of the enterprise's operational resources. Therefore, corporate governance based on the concept of increasing market value enables the enhancement of investment attractiveness, the inflow of financing sources, and the maximization of company value.

New issuers pay a lot of attention to corporate governance. All expenses associated with the implementation of advanced experience in corporate governance can lead to tangible financial benefits in the form of attracting large-scale shareholder capital if the company attracts the interest of strategic and traditional portfolio investors, primarily large pension funds, and insurance companies. The share of institutional investors in capital markets has already reached 50% and is trending upward. Private and institutional investors participate in making investment decisions, and the factors influencing these decisions depend on the type of investor.

Corporate governance is a factor that influences an important component of market value, such as goodwill, which includes prestige, business reputation, clients, and staff of the company. It can be listed on a special account, has no independent market value, and plays a role in mergers and acquisitions. It can be assessed as an intangible asset representing the difference between the sale and book value of the enterprise.

To create a sustainable corporate governance system, it is necessary to:

1. Work on transparency and shareholder engagement. Transparency and openness play a very important role in sustainable management and investor confidence. It is important to engage shareholders and disclose information that allows them to make decisions about investing in sustainable development. The company's ESG goals should be defined and progress reports published, showing both positive and negative outcomes.

2. Incorporate principles of sustainable development into business activities. An organization cannot truly be sustainable if its board of directors does not engage in developing a sustainability strategy. Senior executives should integrate ESG goals into the company's operations and implement mechanisms to monitor their achievement.
3. Appoint a director of sustainable development. Appointing a director or senior manager for sustainable development demonstrates the company's commitment to sustainability principles. This means there is someone to lead "sustainable activities" and provide feedback to the board of directors in this area.
4. Train the board of directors. It is important for board members to not only be aware of the concepts of sustainable development and the ESG agenda but also to understand issues related to climate change, biodiversity, etc. They should also know that all stakeholders play a significant role in sustainable development, from creditors to suppliers.
5. Conduct a sustainable development risk assessment. Risks vary depending on the sector and also act as a catalyst for developing sustainability strategies. Therefore, it is important to spend time assessing these risks in your industry and identifying potential future issues. These could be climate change, resource depletion, social issues, labor rights problems, or any other events that may impact your business.
6. Consider the significance of ESG. Despite ESG being based on "doing the right thing," the material aspect of sustainable development should not be ignored. To encourage the implementation and realization of an ESG strategy, consider linking the sustainable development director's remuneration to the achievement of sustainable goals.

4.4 Business Case: Apple's Integration of Corporate Governance and ESG Investing

The core ideology of a great company remains a guide to action and a source of inspiration. The elements of this ideology are key values and the main purpose. Key values are considered as systems of guiding principles and norms and the main purpose as a global goal defining the meaning of the organization's existence. Key values are fundamental norms of the organization. They remain independent of changes in the external environment. "The key values of Apple are:

1. We offer high-quality products.
2. Our products change people's lives and work.
3. Our products free people from heavy and tedious work, making the world more convenient for living.
4. The quality of our products ensures respect and loyalty from consumers.
5. We are sincerely interested in solving consumer difficulties.
6. We do not compromise ethics for profit.

7. The firm is a corporate citizen of the community.
8. We set "aggressive" goals and make ourselves achieve them.
9. Teamwork is important for the success of the firm.
10. We count on the passion and achievements of each.
11. We support each other and share victories and rewards together.
12. We strive to create an atmosphere where everyone can feel a sense of adventure and joy from working at the firm.
13. Rewards should be moral and financial at the same time.
14. We welcome the interaction of an employee with managers at any level. The attitude of managers to employees is of primary importance.
15. Employees should have reasons to trust the motives and honesty of their superiors".[1]

The management is responsible for creating an environment where the firm's values flourish. These key values allow the company to stand out from the rest; they are not related to external environment conditions but are an internal value of the organization's employees. Apple approaches environmental protection as innovatively as it does to developing its products. To reduce the carbon footprint, we create new projects for solar energy. We switch to safe materials so that products and their production process cause less harm to nature. We protect the forests we use and ensure they are renewed. We even implement more thorough recycling using robots. In its activities, the company is guided by the principles of sustainable development and tries to achieve a balance between socioeconomic and natural-ecological development. The main purpose speaks of why the company exists. It is unchanging, yet it is what inspires the company, stimulating the desire for change. The main purpose should remain unchanged for at least 100 years. The main purpose of Apple is formulated as follows: "Apple aims to offer the best computer technology to students, teachers, creative professionals, and consumers worldwide through its innovative hardware solutions, software, and network applications." Here, the sociological aspect is clearly traced—to bring benefit to society. This component of the vision, the image of the future, is represented by two parts: an ambitious goal and a vivid description. An ambitious goal is something concrete and visible that the company plans to achieve over 10–30 years, a kind of "peak." In 2015, 93% of the energy used by Apple was obtained from renewable sources. But their goal is 100%. Solar panels with a capacity of 32 megawatts, located on 800 roofs, supply energy to the company's facilities in Singapore. Production in China began to reduce the volume of its emissions thanks to a solar energy project with a capacity of 170 megawatts. And data centers around the world operate exclusively on clean energy and provide the transmission of a billion messages, Siri responses, and iTunes music downloads. The vivid description should illustrate what the company will be like if it achieves this goal, i.e., it refers to a time that has not yet come and therefore is related to dreams and aspirations. The visibility of the image is key because only in this way can a distant goal be made visible and convincing to employees. Steve Wozniak, a

[1] https://www.apple.com/

former companion of Jobs, is optimistic about the future of Apple: “The company will continue to thrive,” says the co-founder of Apple. “Jobs was surrounded by the most talented people. And these people are still at Apple. I don’t think the foundation, the core of the company will change with Steve’s departure.” Apple will not burst like a balloon, and its shares will never again be worth less than $400. On September 12, 2023, Apple announced the release of its first “carbon-neutral” product—a new model of the Apple Watch, which leaves no carbon footprint. The carbon neutrality of the Apple Watch line was independently confirmed by SCS Global Services, a leader in environmental standards and certification. In 2019, Apple became the first among more than 400 companies with supplier manufacturers in China to receive the “Master” title from IPE for outstanding indicators of a clean supply chain. “Undoubtedly, the most effective action a supplier can take to address climate change is to switch to renewable energy sources (RES),” says a written statement from Apple. This is why we work closely with suppliers to help them procure more renewable energy sources and together advocate for reliable and economically efficient access to clean electricity in networks around the world. The company added that greenhouse gas emissions from its new iPhone 15 Pro are 28% lower than the company’s 2015 baseline. Apple Corporation (AAPL) announced that it intends to invest $430 billion over the next 5 years in the development and production of its products in the United States. These funds will finance activities such as 5G technology and chip manufacturing.

Apple specifically noted that more than $1 billion of these expenditures will be spent on building a new company building and engineering center in the Research Triangle Park in North Carolina, located near several leading state universities. The company stated that these investments will create at least 3,000 new jobs in various cutting-edge technology fields and provide the state with an economic benefit of more than $1.5 billion.

Other states where Apple intends to expand its presence include Colorado (which is expected to have 700 Apple employees by 2026), Massachusetts, Texas (where the company’s campus is currently under construction), and Iowa, where a new company data processing center (DPC) will be located.

Apple indicated that its recent investments in the American market “significantly exceeded” the 5-year target set by the company in 2018 at $350 billion. The spread of DPCs in the United States expands Apple’s capabilities, allowing the company to reduce content service delays through improved connectivity. Investments in schools and community programs allow the company to increase the loyalty of young consumers to it.

1. How does Apple’s commitment to investing in renewable energy and reducing greenhouse gas emissions from products like the iPhone 15 Pro align with its overall business strategy? Can this approach serve as a model for other tech companies looking to balance profitability with environmental responsibility?
2. What are the potential economic and employment impacts of Apple’s $430 billion investment in the United States, particularly in areas like 5G technology, chip manufacturing, and data processing centers? How might these investments

affect local communities, especially in states like North Carolina, Colorado, Massachusetts, Texas, and Iowa?

4.5 Summary

Corporate governance is not just a set of rules or practices but a fundamental aspect of how a company operates and is perceived in the market. Its importance in investment decisions stems from its significant impact on risk, performance, and reputation. As the business environment continues to evolve, the focus on corporate governance is likely to intensify, further influencing the strategies and choices of investors worldwide.

Corporate governance plays a crucial role in investment decision-making, significantly impacting investor confidence and the long-term success of companies. It encompasses the systems, principles, and processes by which companies are directed and controlled. Good corporate governance is vital for establishing a company's integrity, ensuring effective management, and enhancing shareholder value. It involves various aspects including board structure and practices, executive compensation, shareholder rights, transparency, and accountability.

Investors increasingly recognize that effective corporate governance can lead to enhanced business performance and mitigate risks, making it a critical factor in investment decisions. Companies with strong governance practices are often viewed as more reliable, stable, and better positioned for long-term growth, attracting more investors and potentially leading to a higher valuation in the market.

4.6 Discussion Questions

1. How does effective corporate governance influence investor confidence and decision-making?
2. How do investors evaluate the strength and effectiveness of a company's corporate governance before making investment decisions?
3. In what ways can board diversity influence a company's approach to environmental, social, and governance issues?
4. How can companies balance the need for strong corporate governance with the push for innovative and aggressive sustainability strategies?
5. How do companies measure the impact of shareholder engagement on their sustainability efforts and overall performance?

References

Cambridge Institute for Sustainability Leadership. (2024, November). *What is responsible investment?* https://www.cisl.cam.ac.uk/business-action/sustainable-finance/investment-leaders-group/what-is-responsible-investment

Carroll, A. (1991). Corporate social responsibility evolution of a definitional construct. *Business Society, 38*(3), 268–295. https://doi.org/10.1177/000765039903800303

Harvard Law School Forum on Corporate Governance. (2021, January 24). *ESG disclosures: Guiding principles and best practices for investment managers.* https://corpgov.law.harvard.edu/2021/01/24/esg-disclosures-guiding-principles-and-best-practices-for-investment-managers/

Samans, R., & Jane Nelson, J. (2020). White paper: Integrated corporate governance: A practical guide to stakeholder capitalism for boards of directors. *World Economic Forum.* https://www3.weforum.org/docs/WEF_Integrated_Corporate_Governance_2020.pdf

Chapter 5
ESG Integration in Investment Analysis

Marisa Laokulrach

Abstract The integration of environmental, social, and governance (ESG) factors into investment analysis marks a paradigm shift in the financial landscape. Investors now recognize the significant impact that sustainability factors exert on a company's enduring performance and risk profile, transcending traditional financial metrics. This chapter delves into the manifold dimensions of ESG integration within investment analysis, exploring both fundamental and valuation models.

Fundamental models play a crucial role in unraveling the implications of ESG factors on investment decisions. This comprehensive evaluation ensures that nonfinancial dimensions are integral to the decision-making process, recognizing the intrinsic link between responsible business practices and long-term financial viability. The valuation process now involves meticulous analysis of factors such as climate change exposure, resource management, labor practices, and ethical decision-making. This expanded approach reflects the growing acknowledgment that ESG factors can serve as indicators of a company's intrinsic worth and future financial performance.

The chapter ventures into discerning ESG risks and opportunities embedded in investment decisions. By scrutinizing the relationship between ESG performance and financial outcomes, the exploration is fortified by empirical evidence, presenting a nuanced understanding. Studies consistently underscore a positive correlation between robust ESG performance and enhanced corporate financial performance, particularly pronounced over the long term. This accentuates the substantive contribution of ESG considerations to prolonged value creation.

In essence, the integration of ESG factors into investment analysis signifies a pivotal evolution in the financial landscape. It denotes a departure from conventional metrics, acknowledging the broader spectrum of factors that mold a company's performance and resilience. As investors increasingly seek a holistic understanding of potential investments, ESG integration emerges as a cornerstone,

M. Laokulrach (✉)
International College, National Institute of Development Administration, Bangkok, Thailand
e-mail: marisa.lao@nida.ac.th

M. Bednárová, K. Soratana (eds.), *Environmental, Social, and Governance (ESG) Investment and Reporting*, https://doi.org/10.1007/978-3-031-84235-1_5

aligning financial objectives with sustainable and responsible business practices. This shift underscores a transformative era where considerations beyond mere financial metrics are integral to informed and socially responsible investment decisions.

5.1 Incorporating ESG Factors into Fundamental Analysis and Valuation Models

As environmental, social, and governance (ESG) considerations gain prominence in the investment landscape, the imperative integration of these factors into fundamental analysis becomes increasingly crucial. This section provides a comprehensive overview of the motivations, methodologies, challenges, and benefits associated with incorporating ESG factors into fundamental analysis.

5.1.1 Motivations for Incorporating ESG into Fundamental Analysis and Valuation Models

Investors are increasingly integrating environmental, social, and governance (ESG) factors into their fundamental analysis, driven by different motivations. A key purpose for investors is for long-term value creation. Investors actively search for companies with good ESG practices, perceiving that such entities are better equipped to generate sustained value over extended periods. This expectation holds even in the face of environmental challenges, social controversies, and the evolving landscapes of governance. Companies prioritizing environmental sustainability, social responsibility, and effective governance are perceived as more resilient and better suited to navigate the complexities of the modern business environment. Consequently, investors are drawn to businesses aligning with these values, expecting not only financial returns but also positive contributions to society and the environment.

Another critical motivation for incorporating ESG into fundamental analysis stems from the acknowledgment of ESG factors as crucial indicators of risk. While traditional fundamental analysis predominantly centers on financial metrics, the inclusion of ESG considerations allows investors to delve deeper into a company's risk profile. ESG factors act as early warning signals for potential risks that may not be immediately evident when solely scrutinizing financial statements. Embracing a risk-aware approach that incorporates ESG considerations provides investors with a more comprehensive understanding of a company's overall risk exposure.

The spectrum of environmental risks, encompassing challenges like climate change and resource depletion, social risks tied to labor practices and community relations, and governance risks emanating from issues such as executive compensation and board diversity, forms an integral part of ESG analysis. Investors recognize that neglecting these factors can lead to incomplete risk assessments, potentially exposing portfolios to unforeseen challenges. Consequently, the integration of ESG into fundamental analysis is not merely a trend but a strategic move aimed at enhancing risk management and reinforcing investment decision-making processes.

Beyond risk considerations, shifting investor preferences significantly contributes to the widespread adoption of ESG considerations in fundamental analysis. Investors are no longer singularly focused on financial returns; there is a growing inclination to align investments with ethical, social, and environmental values. This transformative shift in priorities is reshaping the investment landscape, driving an increased demand for responsible and sustainable investment options.

Companies demonstrating a commitment to ESG principles are perceived as more appealing to investors aiming to make a positive impact through their investment decisions. This impact extends beyond financial metrics, encompassing broader societal and environmental implications of a company's operations. By integrating ESG factors into fundamental analysis, investors can align their portfolios with their values, addressing the evolving preferences of a socially conscious investor base.

In conclusion, the motivations behind incorporating ESG factors into fundamental analysis are diverse and compelling. The pursuit of long-term value creation, the recognition of ESG as critical risk indicators, and the evolving preferences of investors toward ethical, social, and environmental considerations collectively drive this trend. As businesses navigate a dynamic landscape, the integration of ESG into fundamental analysis emerges not merely as a tool for enhancing financial performance but as a strategic imperative for mitigating risks, aligning with investor values, and contributing to a more sustainable and responsible global economy.

5.1.2 Methodologies for Integrating ESG into Fundamental Analysis

The incorporation of environmental, social, and governance (ESG) factors into fundamental analysis signals a profound transformation in how investors assess the enduring value and risks associated with their investment portfolios. Driven by a commitment to achieving sustainable returns and proactively managing risks, this evolving approach is reshaping the landscape of investment decision-making. Within this context, we explore the methodologies employed to integrate ESG into fundamental analysis and measure ESG performance within this comprehensive framework.

A primary methodology centers around ESG data integration, serving as the foundational step in this process. This involves assimilating relevant ESG data, encompassing information on a company's environmental impact, social practices, and governance structures. The accuracy and relevance of this ESG data play a pivotal role in conducting precise fundamental analysis, providing investors with a holistic understanding of a company's operational dynamics.

Determining the materiality of specific ESG factors represents a crucial methodology in the integration process. Conducting a materiality assessment becomes imperative for identifying the ESG factors that bear the greatest significance for a particular business. This assessment plays a pivotal role in prioritizing relevant factors, ensuring their centrality in the fundamental analysis process, and aligning them with the unique characteristics of the business under consideration.

Industry benchmarking emerges as a valuable methodology, allowing fundamental analysts to compare a company's ESG performance against that of its industry peers. This comparative analysis offers insights into relative performance and serves as a basis for identifying areas of improvement. By contextualizing a company's ESG metrics within the industry landscape, analysts obtain valuable perspectives on its standing and potential areas for enhancement.

External benchmarks, such as ESG ratings and indices, play integral roles in the integration process. These benchmarks offer a quantitative measure of a company's ESG performance, providing fundamental analysts with valuable reference points to supplement their internal assessments. ESG ratings and indices offer a broader perspective, enabling investors to gauge a company's standing relative to industry benchmarks and assess its performance trajectory over time.

Finally, the incorporation of ESG factors extends beyond quantitative metrics to encompass qualitative analysis. This methodology involves considering a company's environmental concern, social responsibility practices, and governance policies as integral components of its overall quality. By integrating these qualitative aspects, fundamental analysts gain a more comprehensive understanding of a company's ethos and its commitment to responsible business practices.

5.1.3 Integration of ESG Factors into Valuation Analysis

The transformation of environmental, social, and governance (ESG) considerations from mere ethical considerations to integral components in investment decisions and risk evaluations has occurred rapidly. Acknowledging the substantial impact of ESG factors on financial performance, investors now consider their integration into valuation models as indispensable. This in-depth exploration aims to serve as a comprehensive guide for incorporating ESG factors into valuation models, with a specific emphasis on discounted cash flow (DCF). By explanation of adjustments of cash flow projections, discount rates can endeavor to empower investors and analysts with the knowledge necessary to navigate the intricate landscape of ESG-driven valuation.

Discounted Cash Flow (DCF) Model

The discounted cash flow (DCF) model is a foundational tool in financial valuation, aiming to estimate the present value of a company's anticipated future cash flows. The incorporation of ESG considerations into the DCF model necessitates a comprehensive approach that consider various aspects of a company's environmental, social, and governance performance. The DCF model follows a general formula to estimate the present value of future cash flows

$$PV = \frac{CF_1}{(1+r)^1} + \frac{CF_2}{(1+r)^2} + \ldots + \frac{CF_n}{(1+r)^n}$$

where:

- PV is the discounted cash flow, representing the present value of expected future cash flows.
- CF_1, CF_2,…, CF_n are the expected cash flows for each period 1, 2,…, n.
- r is the discount rate, representing the rate of return required by investors. It reflects the time value of money.

The formula calculates the present value of expected cash flows by discounting each cash flow back to its present value using the discount rate. The summation of the discounted cash flows gives the total present value, which represents the estimated intrinsic value of the investment. It's important to note that the DCF model requires careful consideration of factors such as cash flow projections and discount rate selection Additionally, cash flows should be discounted to their present value using the appropriate discount rate for each period.

The ESG performance of companies has the potential to impact operational efficiency, cost management, risk management, and customer perception which ultimately influencing cash flows. Companies that adopt ESG practices not only garner confidence from the financial market and investors but also benefit from a lower cost of capital, ultimately impacting the overall value of the company. The subsequent sections elaborate on how the adoption of ESG practices can lead to adjustments in cash flow projections and the cost of capital.

Adjusting Cash Flow Projections

The integration of environmental, social, and governance (ESG) factors into a company's operational framework has a profound impact on various facets of its performance. Positive ESG practices, such as sustainable operations and ethical supply chain management, play a pivotal role in enhancing operational efficiency, potentially resulting in significant cost savings. Oppositely companies with poor ESG practices may incur increased costs due to regulatory fines or reputational damage.

The refinement of cash flow projections to accommodate ESG factors involves a detailed analysis of how these factors influence revenue, operating expenses, and capital expenditures. This process includes quantifying potential expenses associated with regulatory compliance and integrating cost savings resulting from energy-efficient practices. Through this approach, cash flow projections can encompass the entirety of ESG impacts, providing a more thorough comprehension of a company's financial outlook.

The following example provides the simplicity of how adjusted cash flows can affect company's valuation. Suppose the company is considering a company that involves manufacturing a product. They estimate that implementing environmentally friendly practices in the production process will result in a 10% reduction in energy consumption compared to traditional methods. This reduction in energy consumption is quantified as a cost savings.

Here's a simplified cash flow analysis with the ESG-related performance:

- *Operating Cash Flows*:
  - Annual revenue from the product: \$500,000
  - Traditional production method annual operating costs: \$300,000
  - ESG-friendly production method annual operating costs: \$270,000 (10% cost savings due to reduced energy consumption)
- *Net Cash Flows*:
  - Net cash flow without considering ESG: \$500,000 − \$300,000 = \$200,000 per year
  - Net cash flow with ESG: \$500,000 − \$270,000 = \$230,000 per year
- *Discount Rate*:
  - Assume a discount rate of 10% for simplicity.

Now, let's calculate the present value of cash flows over a 5-year period for both scenarios with the consumption of fixed cash flow for a 5-year period:

PV Without ESG practice:

$$PV_{\text{without ESG}} = \frac{200,000}{(1+0.10)^1} + \frac{200,000}{(1+0.10)^2} + \frac{200,000}{(1+0.10)^3} + \frac{200,000}{(1+0.10)^4} + \frac{200,000}{(1+0.10)^5} = 758,157.85$$

PV With ESG practice:

$$PV_{\text{with ESG}} = \frac{230,000}{(1+0.10)^1} + \frac{230,000}{(1+0.10)^2} + \frac{230,000}{(1+0.10)^3} + \frac{230,000}{(1+0.10)^4} + \frac{230,000}{(1+0.10)^5} = 871,880.96$$

Compare the present values to assess the impact of incorporating the ESG-related performance. If the PV with ESG is higher than the PV without ESG, it suggests that the environmentally friendly practices contribute positively to the project's financial performance. This approach allows companies to consider ESG factors when making investment decisions and assess their impact on long-term financial returns.

A. Operational Efficiency and Cost Management

The influence of ESG factors, categorized under environmental, social, and governance considerations, on operational efficiency and cost management significantly shapes a firm's cash flow estimation. Environmental considerations are pivotal in shaping operational efficiency, with investments in environmentally sustainable practices leading to enhanced efficiency and potential cost savings. Practices such as adopting energy-efficient technologies and implementing waste reduction measures can contribute to substantial cost savings over the long term, positively impacting cash flow. Despite potential upfront costs for eco-friendly technologies, the enduring benefits, including lower energy bills and reduced resource consumption, contribute to overall cost containment.

Social factors also play a crucial role in operational efficiency and cost management. Socially responsible practices, such as implementing employee training programs and diversity initiatives, have the potential to elevate workforce productivity and satisfaction. A motivated and well-trained workforce tends to be more efficient, positively impacting operational performance and, consequently, cash flow. Investments in employee well-being, although incurring initial costs, can result in lower turnover rates, reduced recruitment expenses, and higher employee productivity over time. The cumulative effect of these social considerations contributes to cost savings and exerts a positive influence on cash flow.

Governance factors are integral to both operational efficiency and cost management. Strong governance practices enhance decision-making processes and organizational efficiency. Efficient governance structures contribute to streamlined operations, reducing bureaucratic costs and positively impacting cash flow. Additionally, effective governance plays a crucial role in cost management by helping companies avoid legal issues, regulatory fines, and other governance-related costs. The minimizing legal risks can help a company to safeguard its cash flow from unexpected outflows due to non-compliance or legal challenges. The governance factors become instrumental in preserving the financial stability and resilience of the firm's cash flow.

B. Regulatory Compliance

Compliance with environmental regulations is paramount for companies aiming to evade fines and legal penalties. Proactively managing and reducing environmental impact minimizes compliance issues, safeguarding cash flow from unexpected regulatory costs. Adhering to social regulations, encompassing standards related to labor and human rights, acts as a mitigation strategy against legal risks. Companies upholding fair labor practices and fostering positive workplace environments contribute to regulatory compliance, averting potential legal expenses.

Strong governance practices, characterized by adherence to corporate laws and regulations, play a crucial role in helping companies navigate legal pitfalls. This commitment to governance fosters regulatory compliance, shielding cash flow against legal challenges and associated costs. Integrating ESG considerations into regulatory compliance strategies ensures a holistic approach, covering environmental, social, and governance factors, thereby contributing to the overall financial resilience of the company.

C. Risk Management

Proactively managing environmental risks, such as those linked to climate change or pollution, serves to prevent operational disruptions and associated costs. This risk mitigation strategy fosters stable operations and, consequently, positive cash flow. Addressing social risks, including issues within the supply chain or community relations, acts as a preventive measure against reputational damage and potential legal actions. Effective risk management in social aspects protects the firm's cash flow from unforeseen liabilities.

Strong governance practices are inherently linked to effective risk management. Transparent and accountable governance structures enable companies to identify and address risks more efficiently, protecting cash flow from unexpected shocks. By integrating ESG considerations into risk management frameworks, companies can assess and mitigate risks associated with environmental, social, and governance factors, ensuring a more resilient and financially sound operation.

D. Market Opportunities

With increasing environmental awareness, there is a growing market demand for sustainable products and services. Companies aligning their offerings with environmental considerations can tap into new market opportunities, potentially leading to increased sales and positive cash flow effects. Consumers are placing higher value on socially responsible businesses. Companies demonstrating a commitment to social issues, such as diversity in labor forces and fair labor practices, may attract a broader customer base. This contributes to revenue growth and positively influences cash flow.

Companies with good governance practices may be viewed more favorably by investors and customers alike. The trust generated by good governance can open doors to new market opportunities, partnerships, and business relationships, positively impacting cash flow. As markets evolve to prioritize sustainability and social responsibility, companies that integrate ESG considerations into their business strategies are better positioned to seize emerging market opportunities, enhancing their financial performance and positively impacting cash flow.

E. Supply Chain Resilience

Enhancing supply chain resilience involves assessing and addressing environmental risks, such as disruptions related to climate change or resource scarcity. This strategic approach contributes to a more robust supply chain that can withstand challenges, ensuring a continuous flow of goods and services. Importantly, it protects cash flow by preventing potential disruptions that could adversely affect a company's operations.

The evaluation and management of social risks in the supply chain, such as labor disputes or human rights violations, play a pivotal role in establishing a stable and ethical supply chain. By proactively addressing social controversies and disruptions, a company can safeguard its cash flow. This approach ensures a supply chain that aligns with ethical principles and fosters long-term sustainability.

Governance practices within the supply chain, characterized by transparency and accountability, significantly contribute to better risk management. Effective governance serves as a preventive measure against issues like corruption or unethical practices in the supply chain. This not only protects cash flow but also fortifies the overall integrity and reliability of the supply chain. As companies increasingly recognize the interconnectedness of their operations with the broader supply chain, integrating ESG factors into supply chain management becomes essential for maintaining resilience and protecting cash flow.

F. Consumer Perception

With increasing consumer concern about environmental issues, companies that demonstrate a commitment to sustainability and eco-friendly practices can enhance their brand reputation among environmentally conscious consumers. Positive consumer perception translates into increased sales and brand loyalty, thus positively impacting cash flow. Socially responsible and ethical business practices, such as labor fairness practices and community engagement, contribute to building a good brand image. Consumers tend to prefer companies with a strong social conscience, influencing their purchasing decisions and, consequently, positively impacting cash flow.

Encompassing transparent decision-making, ethical business conduct, and effective risk management can enhance a company's credibility. A reputation for good governance positively influences brand perception, attracting customers who value accountability and integrity. This, in turn, has a positive impact on cash flow as consumers align with companies that uphold high governance standards. As consumer preferences evolve to prioritize ethical and sustainable business practices, companies that integrate ESG considerations into their brand strategy are better positioned to build trust, enhance their market share, and positively impact cash flow.

Discount Rate Adjustments

Before exploring the specific impact of environmental, social, and governance (ESG) factors, it is essential to grasp the role of the discount rate in valuation models. Discount rates, expressed as the cost of equity or a weighted average cost of capital (WACC), indicate the risk associated with an investment and are used to discount future cash flows for intrinsic value determination.

The discount rate is a critical parameter in calculating the present value of future cash flows, reflecting the time value of money and the risk associated with an investment. Traditionally determined based on financial metrics, the discount rate is evolving as non-financial considerations gain prominence.

ESG factors can significantly influence this risk profile. Companies with strong ESG practices may be deemed a lower risk, justifying a lower discount rate. Conversely, poor ESG performance may elevate perceived risks, leading to a higher discount rate. Adjusting the discount rate based on ESG performance ensures that the model accurately reflects the risk associated with the company's cash flows, influencing the present value of future cash flows.

A. ESG and Risk Perception

The integration of environmental risks into ESG considerations acknowledges the profound impact of factors such as climate change impacts, regulatory changes, and resource scarcity on a company's operations. These risks pose direct threats to a company's future cash flows, influencing the perceived risk in valuation models. Investors, cognizant of the potential disruptions posed by environmental challenges, consider these risks when determining the appropriate discount rate. The assessment of a company's ability to navigate and adapt to environmental risks becomes a critical factor in shaping investor perceptions and prompting adjustments in discount rates.

Social considerations within the ESG framework encompass a broad spectrum, including labor practices, community relations, and product safety. The risks associated with these social factors go beyond immediate financial implications and may result in litigation, regulatory fines, or reputational damage. Evaluating social risks is crucial for understanding a company's overall risk profile. When calculating discount rates, investors consider these social risks, understanding that issues like labor disputes or product safety scandals can have profound consequences. The perceived risk in discount rate calculations reflects investors' assessment of a company's ability to effectively manage and mitigate social risks.

Governance risks within ESG considerations relate to issues such as executive compensation, adherence to ethical practices, and the effectiveness of corporate governance structures. Weak governance practices can lead to management inefficiencies, legal complications, and a lack of transparency. Investors adjust discount rates to accommodate governance-related risks, recognizing that robust governance is vital for long-term stability. Integrating governance risks into risk perception reflects investors' evaluation of a company's leadership, ethical standards, and overall governance framework.

B. Risk Premiums

ESG-related risks, ranging from environmental challenges to social controversies, are quantified through risk premiums applied to discount rates. This approach recognizes that companies face additional uncertainties and potential challenges associated with their ESG posture. As scrutiny regarding ESG practices intensifies, the quantification of risk premiums becomes integral to discount rate adjustments, offering a more precise evaluation of the specific challenges a company may encounter.

Environmental risks, encompassing climate change impacts, regulatory changes, and resource scarcity, are translated into risk premiums. Companies with higher exposure to environmental risks may attract higher-risk premiums, leading to an upward adjustment in the discount rate. Investors seek compensation for uncertainties related to a company's ability to mitigate environmental challenges and adapt to a rapidly changing ecological landscape.

Social factors, including labor practices, community relations, and product safety, introduce a layer of complexity beyond immediate financial implications. Litigation, regulatory fines, or reputational damage arising from social risks contributes to risk premiums. The assessment of these risks factors into discount rate adjustments, recognizing the broader consequences that social controversies may have on a company's financial outlook.

Governance risks, covering issues like executive compensation and adherence to ethical practices, introduce management inefficiencies and legal complications. The quantification of risk premiums acknowledges the governance-related challenges a company may face, influencing discount rate adjustments. Investors recognize that weak governance structures pose additional risks that warrant compensation through higher-risk premiums.

C. Cost of Equity Adjustment

The cost of equity, a pivotal component of the discount rate, reflects the return investors require for bearing the risks associated with a company's equity. ESG considerations significantly influence the cost of equity, shaping the overall cost of capital for companies.

Companies with robust ESG practices are perceived as less risky by investors, leading to a lower cost of equity. This adjustment reflects the understanding that sustainable and ethical business practices contribute to long-term stability. Conversely, companies facing significant ESG-related risks may incur a higher cost of equity, reflecting the additional risk investors associate with their operational practices.

The adjustment of the cost of equity based on ESG considerations underscores the interplay between sustainability and capital structure. Companies integrating ESG practices into their strategic framework find themselves not only aligning with investor expectations but also influencing the very core of their capital structure. As the cost of equity adjusts to accommodate ESG-related factors, it becomes clear that sustainability is not merely a peripheral concern but an integral determinant of a company's overall financial architecture.

D. Weighted Average Cost of Capital (WACC)

The adjustment of discount rates based on ESG considerations extends beyond the cost of equity to the comprehensive evaluation encapsulated by the weighted average cost of capital (WACC). WACC considers both the cost of debt and the cost of equity, providing a holistic perspective that mirrors the interconnected nature of financial and sustainability considerations.

Companies actively managing ESG risks may secure lower borrowing costs, positively impacting the cost of debt. Lenders and bondholders increasingly factor a company's ESG performance into assessments of creditworthiness. The adjustment of the cost of debt reflects not only the financial prudence of ESG practices but also the acknowledgment of their role in mitigating risks associated with debt obligations.

The integration of ESG factors into WACC calculations highlights the evolving criteria for assessing creditworthiness. Companies aligning their financial strategies with effective ESG practices may experience a more favorable WACC, influencing their overall cost of capital. This integration acknowledges that sustainability is not an ancillary consideration but a fundamental factor shaping a company's financial landscape.

E. Influence of ESG Ratings or Indices: External Benchmarks Shaping Perceptions

External benchmarks, such as ESG ratings and inclusion in ESG-focused indices, play a crucial role in shaping investor perceptions and influencing discount rate adjustments. These benchmarks serve as reference points, signifying a company's commitment to sustainable and responsible business practices.

High ESG ratings, indicative of strong environmental, social, and governance performance, result in lower perceived risk. Investors often rely on external ESG ratings as a measure of excellence in sustainability practices. The integration of high ESG ratings into risk perception influences discount rate adjustments, reinforcing the notion that companies excelling in ESG considerations are better positioned for long-term success.

Inclusion in ESG-focused indices further amplifies a company's alignment with global ESG standards. These indices represent a collective acknowledgment of a company's commitment to meeting stringent sustainability benchmarks. Investors, recognizing the global significance of these indices, incorporate inclusion as a factor in discount rate adjustments. The influence of ESG indices goes beyond individual company evaluations, shaping a broader understanding of sustainability as a critical component of investment analysis.

5.2 Identifying Risks and Opportunities in Investment Decision-Making

In the contemporary landscape of investments, environmental, social, and governance (ESG) factors have evolved into pivotal considerations, not only influencing ethical investment practices but also shaping financial performance. This part explores the complexities of identifying ESG risks and opportunities in investment decision-making, examining methodologies, challenges, and the transformative impact of integrating ESG considerations into the investment process.

5.2.1 Identifying Environmental Risks and Opportunities

Assessing environmental risks and opportunities is pivotal for responsible investing and sustainable business practices. Understanding a company's susceptibility to environmental challenges provides investors with crucial insights into potential risks impacting its operations and opportunities for sustainable growth. Within this framework, three critical aspects of identifying environmental risks and opportunities are examined: climate change exposure, resource management, and regulatory compliance.

Climate Change Exposure

The evaluation of a company's exposure to climate change entails an assessment of its vulnerability to physical risks. These risks encompass the potential impacts of extreme weather events, sea-level rise, and other climate-related phenomena. Companies operating in regions prone to hurricanes, floods, or wildfires may encounter heightened risks affecting their infrastructure, supply chains, and overall operational resilience. Identifying these physical risks enables investors to understand the potential implications on a company's assets and operations.

In addition to physical risks, companies are exposed to transition risks associated with shifts towards a low-carbon economy. Regulatory changes aimed at reducing carbon emissions may impact certain industries and business models. Investors need to identify companies that are proactively transitioning to more sustainable practices, as they may present long-term opportunities. Businesses embracing renewable energy sources, implementing energy-efficient technologies, or participating in carbon offset initiatives are examples of companies navigating transition risks and leveraging opportunities.

Resource Management

Supply Chain Disruptions: Companies heavily reliant on scarce resources, such as water, minerals, or energy, face risks related to supply chain disruptions. Identifying these risks involves evaluating a company's dependence on critical resources and understanding how disruptions can impact its operations. For example, a manufacturing company dependent on a specific mineral may face challenges if geopolitical events disrupt the supply chain for that mineral. Investors assessing resource management risks aim to understand how well a company diversifies its supply sources and its resilience to potential disruptions.

Efficient resource usage is crucial for sustainable business practices. Companies that are inefficient in their resource utilization may face increased costs as resources become scarcer. Identifying resource management risks involves evaluating a company's strategies for mitigating resource-related challenges. This may include efforts to reduce waste, optimize production processes, and adopt circular economy practices. Investors looking for opportunities may favor companies that demonstrate a commitment to sustainable resource management, as efficient use of resources can contribute to cost savings and long-term viability.

Regulatory Compliance

Regulatory changes related to environmental practices can have a profound impact on companies, influencing their operations and financial performance. Identifying risks associated with regulatory compliance involves monitoring current regulations and anticipating potential future changes. For example, stricter emissions standards, waste disposal regulations, or requirements for environmental impact assessments can affect businesses across various sectors. Investors seeking to manage environmental risks may prioritize companies that proactively adapt to existing regulations and are prepared for potential future changes.

Companies that actively engage in proactive compliance and adaptation strategies may present opportunities for investment. This includes implementing environmentally friendly practices, investing in technologies that reduce environmental impacts, and participating in voluntary sustainability initiatives. For instance, a company adopting renewable energy sources or incorporating circular economy principles into its business model may position itself as a leader in environmental stewardship. Investors recognizing such initiatives as indicators of responsible corporate behavior may find opportunities for sustainable and socially responsible investments.

5.2.2 *Identifying Social Risks and Opportunities*

One of the fundamental aspects of socially responsible investing involves evaluating a company's social practices. This assessment encompasses various dimensions, including labor practices, community relations, and diversity and inclusion. Understanding how companies manage these social aspects is crucial for investors aiming to make informed decisions aligned with ethical and sustainable principles.

Evaluating a company's labor practices involves a comprehensive analysis of its approach to employee well-being, working conditions, fair wages, and adherence to labor laws. Social risks may arise if a company is associated with controversies related to labor practices, such as poor working conditions or unfair compensation. On the flip side, there are opportunities for companies that prioritize creating positive workplace environments.

Labor Practices

Companies with inadequate labor practices may face significant social risks. Instances of labor exploitation, unsafe working conditions, or violation of labor laws can lead to reputational damage, legal challenges, and increased scrutiny from stakeholders. Investors need to be vigilant in identifying companies with poor labor practices, as these issues can have long-lasting consequences for both the affected workers and the company's overall reputation.

Conversely, companies that invest in creating positive workplace environments present opportunities for investors. Organizations that prioritize fair wages, employee well-being, and compliance with labor laws are likely to attract and retain talent. Positive labor practices contribute to a motivated and satisfied workforce, which can enhance productivity and innovation. Investing in companies committed to ethical labor practices aligns with social responsibility and may lead to long-term sustainability.

Community Relations

A company's relationship with local communities is another critical aspect of social responsibility. Poor community relations can expose companies to social risks, including reputational damage, regulatory scrutiny, and challenges in obtaining community support for business activities. Companies that neglect or harm local communities may face social risks. Negative impacts on communities, such as environmental degradation, displacement of residents, or disregard for local cultures, can lead to protests, legal actions, and reputational harm. Investors must assess a company's community engagement practices to identify potential risks associated with its impact on local populations.

Positive community relations present opportunities for companies to build trust and goodwill. Engaging with local communities transparently and respectfully can enhance a company's reputation, mitigate potential conflicts, and contribute to sustainable development. Investors recognizing and supporting companies with strong community relations practices contribute to fostering responsible corporate behavior.

Diversity and Inclusion

Diversity and inclusion are increasingly recognized as essential components of responsible corporate practices. Evaluating a company's approach to diversity involves assessing its commitment to creating an inclusive workforce that values individuals from diverse backgrounds. Companies with insufficient commitment to diversity and inclusion may face social risks. Workplace discrimination, lack of equal opportunities, and biased practices can lead to negative perceptions and potential legal challenges. Investors need to consider a company's stance on diversity and inclusion to identify potential risks related to its organizational culture.

Promoting diversity and inclusion presents opportunities for companies to increase innovation and improve decision-making. Research result suggests that diverse teams contribute to a broader range of perspectives, creativity, and problem-solving capabilities. Investors recognizing the importance of diversity may find opportunities in companies actively championing inclusive practices.

5.2.3 Identifying Governance Risks and Opportunities

Corporate governance serves as the framework through which companies make decisions, and its effectiveness is crucial for mitigating risks and identifying investment opportunities. Three key aspects of governance—board effectiveness, executive compensation, and ethical decision-making—play a pivotal role in shaping a company's overall governance structure.

Broad Effectiveness

Governance risks may emerge from boards that lack effectiveness in providing proper oversight of a company's operations. Identifying these risks involves a thorough evaluation of the composition, independence, and expertise of the company's board. Ineffective boards can pose governance risks by failing to exercise appropriate oversight, leading to mismanagement, conflicts of interest, and strategic missteps. Boards lacking independence or expertise in critical areas may struggle to make informed decisions, exposing the company to vulnerabilities. Investors need to scrutinize board structures to identify any governance risks associated with inadequate oversight and decision-making.

Opportunities, on the other hand, are presented by companies that boast strong governance structures promoting transparency and accountability. Companies with strong governance structures present opportunities for investors. These are characterized by boards with diverse expertise, independence, and effective oversight mechanisms. Transparent decision-making processes, clear delineation of responsibilities, and a commitment to ethical standards contribute to the effectiveness of the board. Investing in companies with robust governance practices aligns with principles of responsible investing and contributes to long-term sustainability.

Executive Compensation

The structure and amount of executive compensation can be a governance risk if not aligned with long-term shareholder value. Identifying these risks involves assessing the transparency and performance linkage of executive compensation practices. Excessive or poorly structured executive compensation can create governance risks. Compensation that is not linked to long-term shareholder value may incentivize short-term decision-making at the expense of sustainable growth. Investors should evaluate whether executive compensation structures align with the company's overall performance and shareholder interests.

Companies with transparent and performance-linked executive compensation practices present opportunities for investors. Aligning executive pay with long-term performance metrics and shareholder value ensures that executives are incentivized to make decisions that contribute to the company's sustained success. Such practices foster a culture of accountability and responsible governance, making these companies attractive investment opportunities for those prioritizing ethical and sustainable practices.

Ethical Decision-Making

Instances of ethical misconduct, such as fraud, corruption, or unethical business practices, can present notable governance risks. These transgressions may result in legal repercussions, harm to reputation, and the erosion of shareholder trust. Evaluating a company's ethical track record and dedication to responsible business conduct is crucial for investors to identify potential governance risks.

Companies that prioritize a robust ethical culture present appealing prospects for investors. Nurturing a commitment to ethical decision-making and responsible business practices contributes to long-term sustainability. Ethically conscious companies are better positioned to navigate challenges, establish trust with stakeholders, and uphold a positive reputation. Investing in companies with a demonstrated history of ethical governance aligns with the principles of responsible investing.

5.3 The Connection Between ESG Performance and Investment Performance

In recent times, the intersection of environmental, social, and governance (ESG) factors with investment decisions has garnered significant attention. Investors now acknowledge that a company's ESG performance plays a pivotal role in determining its long-term sustainability and, by extension, its investment performance. This section aims to explore the nuanced connection between ESG performance and investment outcomes, examining how environmental stewardship, social impact, and effective governance practices can either enhance or diminish the financial success of an investment.

5.3.1 ESG Integration in Investment Strategies

The incorporation of ESG considerations into investment strategies has emerged as a prominent practice, signaling a paradigmatic change in the investment landscape. Recognizing nonfinancial factors as crucial determinants of investment outcomes underscores a dedication to sustainability and responsible business practices. This examination explores diverse approaches to ESG integration, emphasizing the importance of aligning investment portfolios with principles that advocate for environmental stewardship, social welfare, and robust governance.

Negative and Positive Screening

Negative screening represents a risk mitigation strategy where companies with insufficient ESG performance or involvement in controversial activities are excluded from the investment universe. Investors adopting negative screening aim to protect their portfolios from potential regulatory, reputational, or operational challenges associated with companies lacking in ESG practices. This approach reflects a dedication to ethical investing by avoiding companies engaging in practices inconsistent with responsible business conduct.

In contrast, positive screening is a proactive strategy involving the active selection of companies with robust ESG credentials for inclusion in investment portfolios. This approach aligns with the philosophy of supporting companies that demonstrate responsible business practices, environmental awareness, and a commitment to social welfare. Positive screening seeks to identify ESG leaders, offering investors the opportunity to contribute to the success of companies with sustainable and ethical practices.

ESG Integration in Fundamental Analysis

ESG integration in fundamental analysis signifies an approach that integrates ESG factors into traditional financial analysis models. Analysts assess a company's ESG performance in conjunction with conventional financial metrics to obtain a comprehensive understanding of its overall risk profile and growth potential. This integrated analysis seeks to offer investors a more holistic view of a company's intrinsic value, acknowledging the interconnected nature of financial and nonfinancial considerations.

Thematic Investing

Thematic investing focuses on specific ESG themes, such as clean energy, sustainable agriculture, or social impact. Investors allocate capital to companies that contribute positively to these themes, aiming to support industries aligned with sustainable development goals and societal needs. Thematic investing allows investors to target their financial resources toward sectors that address pressing global challenges, fostering positive change through strategic investments.

Impact Investing

Going beyond conventional financial returns, impact investing prioritizes investments that generate measurable positive social and environmental impacts. Investors actively seek opportunities to contribute to solutions for global challenges, such as climate change, poverty, or inequality. Impact investing represents a conscious effort to align investment strategies with broader societal goals, emphasizing the potential for financial returns alongside meaningful contributions to global well-being.

Engagement and Active Ownership

ESG integration involves active engagement with invested companies and the exercise of ownership rights to influence positive change. Shareholders collaborate with companies to improve ESG practices, enhance disclosure, and align business strategies with sustainable principles. Active ownership reflects a commitment to fostering positive corporate behavior and acknowledges the role investors can play in steering companies toward more responsible practices.

Risk-Return Optimization

Recognizing the intricate relationship between risk and return, ESG integration aims to optimize risk-return profiles by factoring in ESG considerations. Investors assess how a company's ESG performance may impact its ability to navigate challenges, seize opportunities, and deliver sustainable financial returns over the long term. This approach reflects a holistic view of risk that encompasses both traditional financial metrics and nonfinancial ESG factors.

ESG Index Investing

Investors may choose ESG index investing, where portfolios mirror the composition of established ESG indices such as the MSCI ESG Leaders Index, the S&P 500 ESG Index, and DJSI. ESG indices select companies based on their ESG performance, providing investors with a passive investment strategy aligned with recognized ESG benchmarks. This approach enables investors to track the performance of companies meeting specific ESG criteria, offering a systematic and benchmarked approach to ESG integration.

The integration of ESG factors in investment strategies reflects a broader acknowledgment of the multifaceted nature of investment decisions. By incorporating environmental, social, and governance factors, investors seek to align their portfolios with sustainable and responsible business practices. The diverse approaches discussed—encompassing negative and positive screening, thematic investing, impact investing, engagement, and ESG index investing—underscore the adaptability and versatility of ESG integration strategies. As responsible investing gains momentum, the incorporation of ESG considerations is poised to become an integral aspect of investment decision-making, contributing to a more sustainable and ethical financial landscape.

5.3.2 Impacts of ESG Performance on Financial Performance

The influence of environmental, social, and governance (ESG) performance on financial outcomes is in different perspectives, emphasizing the growing recognition that nonfinancial factors play a crucial role in determining a company's overall prosperity. The integration of ESG considerations into business strategies has far-reaching implications across various dimensions of financial performance.

One fundamental aspect where ESG considerations impact financial performance is through risk mitigation and long-term value creation. Companies excelling in ESG performance showcase superior capabilities in identifying and managing risks associated with environmental disasters, social controversies, and governance lapses. This risk mitigation strategy contributes to the company's fiscal well-being, fostering sustained long-term value creation. Numerous studies consistently highlight a positive correlation between high ESG performance and enhanced corporate financial performance. This correlation is particularly evident over the long term, suggesting that ESG considerations actively contribute to prolonged value creation.

Moreover, ESG-driven initiatives contribute to cost reduction and operational efficiency. Companies that prioritize ESG factors often implement sustainable practices, such as energy-efficient initiatives, waste reduction programs, and sustainable supply chain management. The subsequent decrease in operational costs enhances profitability, positively impacting financial performance. A real-world example is Unilever's Sustainable Living Plan, which not only enhances ESG performance but also contributes to operational efficiency and financial gains.

Access to capital and a lower cost of capital represent additional dimensions of the financial impact of ESG performance. Investors increasingly recognize the financial implications of ESG performance, and companies with robust ESG credentials often find it easier to access capital, often at a lower cost. This is because companies with strong ESG practices are perceived as lower-risk entities, attracting capital from ESG-focused investors and sustainable finance instruments. Financial instruments like green bonds, sustainability-linked loans, and other forms of green financing have gained traction, allowing companies dedicated to ESG principles to unlock capital for sustainable projects and initiatives.

Enhanced investor relations and market recognition are also influenced by a company's ESG performance. ESG performance is becoming a pivotal factor in investor decision-making, especially for those inclined toward responsible investing. Investors are drawn to companies demonstrating a commitment to sustainability and ethical practices, and this heightened investor interest positively impacts a company's stock performance and market recognition. Companies excelling in ESG performance are often included in ESG indices like the Dow Jones Sustainability Index (DJSI) or the FTSE4Good Index, signifying not only market recognition but also attracting investments from funds tracking these indices.

The impact of ESG performance extends beyond financial markets to brand reputation and consumer loyalty. ESG performance significantly influences brand reputation and consumer loyalty as consumers become more socially and environmentally conscious. Consumers increasingly prefer products and services from companies with strong ESG commitments, translating into increased market share and improved financial performance. Studies indicate that a significant percentage of consumers consider a company's social and environmental commitments when making purchasing decisions, providing a competitive advantage to brands aligning with consumer values on sustainability.

The integration of ESG considerations into business strategies has transformative implications for financial performance. From risk mitigation and long-term value creation to cost reduction, access to capital, investor relations, and consumer loyalty, companies prioritizing ESG factors stand to benefit across various dimensions. As global awareness of sustainability grows, the financial landscape is evolving to underscore the importance of responsible business practices, making ESG considerations not just a moral imperative but a strategic necessity for companies seeking long-term financial success.

5.3.3 Empirical Evidence on Connection Between ESG Performance and Financial Performance

In recent years, there have been extensive academic and professional studies on the intricate relationship between environmental, social, and governance (ESG) factors and corporate outcomes, shedding light on their multifaceted impact on financial performance. The results from studies are diverse, reflecting the complexity and multidimensional metrics of ESG and financial performance.

Loof and Stephan (2019) brought forth compelling evidence, revealing that companies with higher ESG scores not only exhibit increased profitability and stock value but also enjoy higher returns and encounter lower risks in merger and acquisition transactions. Friede et al.'s (2015) study provided additional insights, suggesting that companies incorporating ESG elements into their strategies demonstrate lower stock performance volatility compared to their industry counterparts. Building on this, Kumar et al. (2016) employed a novel quantitative model to assess the returns and volatility of stocks in the Dow Jones Sustainability Index (DJSI), uncovering that companies embracing ESG measures outperform their peers by showcasing lower volatility and higher risk-adjusted returns.

Khan et al. (2016) identified the realm of corporate social responsibility (CSR) activities and their association with financial outcomes. Their findings established a positive correlation, indicating that companies with robust CSR practices tend to exhibit higher profitability and lower risk. Luo and Bhattacharya (2006) identified that firms with strong corporate social responsibility tend to gain higher customer satisfaction and lead to better financial returns. In Thailand, Laokulrach's (2022) study further affirmed the positive impact of ESG adoption, revealing that companies in six out of seven industries, that embraced ESG principles, delivered superior risk-adjusted returns compared to industry peers.

The research landscape broadened with Przychodzen and Przychodzen's (2013) examination of 85 S&P 500 companies that integrated sustainability into their corporate strategies reinforced this perspective by showcasing that such companies can generate abnormal returns with lower risk. The insights extended globally as Zhang et al. (2018) discovered that Chinese companies frequently releasing sustainability-related news tend to generate higher stock returns. On the Brazilian Stock Market, Azevedo et al. (2016) found that sustainable companies demonstrate greater returns compared to their counterparts, offering a glimpse into the global relevance of ESG considerations.

However, some studies have shown a negative relationship between ESG performance and financial performance identifying the costs incurring in ESG implementation might outweigh their benefits. The narrative takes a nuanced turn with Meher,

Hawaldar, Mohapatra, Spulbar, and Birau's (2020) study on Indian companies, which identified a negative connection between environmental and good governance adoption and returns, coupled with a negative relationship with volatility. Similarly, Brammer et al.'s (2006) findings added complexity, revealing that a set of environmental and social factors, including environment, employment, and community activities, surprisingly exhibit a negative relationship with returns. This identifies the costs of CSR that might be incurred without immediate benefits. Bauer et al.'s (2005) analysis of an international database with 103 ethical mutual funds in Germany, the UK, and the USA provided a contrasting perspective by finding no significant differences in risk-adjusted returns between ethical and conventional funds from 1990 to 2001 when examining the investment style.

The influence of ESG on financial performance also varies across industries or over time horizons. Industries with higher sensitivity to environmental regulations might experience a stronger positive effect of ESG on their financial outcomes. Eccles et al.'s (2014) exploration highlighted that in long term, high-sustainability companies significantly outperformed the others, particularly in industries where environmental and social issues are the main focal points. Nollet et al. (2016) found the long-term impact of ESG practices on financial performance.

These diverse findings underscore the intricate interplay between ESG factors and financial performance, hinting at the contextual difference and industry-specific dynamics that shape the outcomes. While some studies emphasize the positive correlation between ESG practices and financial success, others reveal complexities, highlighting the need for a nuanced understanding of how ESG considerations manifest in different business environments. The variability of results can often be caused by differences in methodology, industry characteristics, and time horizons. Overall, the prevailing results lean toward a positive or at least non-negative relationship, identifying that ESG integration can be aligned with financial performance, especially in the longer term. This understanding pinpoints the importance of context consideration in evaluating the financial implications of ESG investment. The global nature of these studies also emphasizes the universality of the discourse, signaling that the impact of ESG considerations on financial outcomes is a multifaceted and evolving area of research with implications for companies, investors, and policymakers worldwide.

5.4 Summary

The incorporation of environmental, social, and governance (ESG) considerations into investment strategies represents a profound shift in the financial landscape, driven by the pursuit of long-term value creation, risk mitigation, and evolving investor preferences. Various approaches, ranging from negative screening to impact investing, underscore the diverse ways in which ESG integration manifests. Despite its transformative potential, challenges such as standardization issues, greenwashing concerns, and the delicate balance between financial and ESG objectives persist.

Environmental risks and opportunities entail the assessment of climate change exposure, resource management, and regulatory compliance. Social risks and opportunities involve evaluating labor practices, community relations, and diversity and inclusion. Governance risks and opportunities encompass scrutinizing board effectiveness, executive compensation, and ethical decision-making. Identifying and understanding these factors empower investors to make informed decisions aligned with ESG considerations, fostering sustainable and responsible investment practices.

The growing recognition of ESG factors in investment decisions is evidenced by various strategies, including negative and positive screening, thematic investing, and engagement. The multifaceted impacts of ESG performance on financial outcomes, including risk mitigation, long-term value creation, and enhanced operational efficiency, highlight its integral role in shaping corporate financial success. Access to capital, lower costs, improved investor relations, and market recognition further validate the significance of robust ESG practices. As ESG considerations continue to evolve, a nuanced understanding of their integration into investment practices becomes imperative for navigating the complexities of the modern financial landscape.

5.5 Discussion Questions

1. What are the key challenges in incorporating ESG factors into investment analysis, particularly in terms of standardization and data quality?
2. How does the consideration of ESG factors affect the determination of the discount rate used in valuation models?
3. Discuss the relationship between a company's ESG performance and its access to capital. How does improved access to capital influence cash flow projections?
4. How has the integration of ESG considerations into investment strategies evolved over recent years, and what motivates this shift in the financial landscape?
5. Discuss how sustainable practices driven by ESG considerations, such as energy efficiency and waste reduction, lead to cost reduction and operational efficiency.

References

Azevedo, V. G., Santos, A. A., & Campos, L. M. (2016). Corporate sustainability and asset pricing MODELS: Empirical evidence for the Brazilian stock market. *Production, 26*(3), 516–526. https://doi.org/10.1590/0103-6513.201115

Bauer, R., Koedijk, K., & Otten, R. (2005). International evidence on ethical mutual fund performance and investment style. *Journal of Banking and Finance, 29*, 1751–1767. https://doi.org/10.1016/j.jbankfin.2004.06.035

Brammer, S., Brooks, C., & Pavelin, S. (2006). Corporate social performance and stock returns: UK evidence from disaggregate measures. *Financial Management, 35*(3), 97–116. https://doi.org/10.2469/dig.v37.n2.4619

Eccles, R. G., Ioannou, I., & Serafeim, G. (2014). The impact of corporate sustainability on organizational processes and performance. *Management Science, 60*(11), 2835–2857.

Friede, G., Busch, T., & Bassen, A. (2015). ESG and financial performance: Aggregated evidence from more than 200 empirical studies. *Journal of Sustainable Finance and Investment, 5*(4), 210–333.

Khan, M., Serafeim, G., & Yoon, A. (2016). Corporate sustainability: First evidence on materiality. *The Accounting Review, 91*(6), 1697–1724. http://www.jstor.org/stable/24907173

Kumar, A. N., Smith, C., Badis, L., Wang, N., Ambrosy, P., & Tavares, R. (2016). ESG factors and risk-adjusted performance: A new quantitative model. *Journal of Sustainable Finance & Investment, 6*(4), 292–300. https://doi.org/10.1080/20430795.2016.1234909

Laokulrach, M. (2022). The influence of sustainable development on stock return and volatility in Thailand's stock exchange during the COVID-19 pandemic. *Asian Economic and Financial Review, 12*(9), 751–765. https://doi.org/10.55493/5002.v12i9.4592

Loof, H., & Stephan, A. (2019). *The impact of ESG on stocks' downside risk and risk adjusted return* (Vol. 477). Working Paper Series in Economics and Institutions of Innovation.

Luo, X., & Bhattacharya, C. B. (2006). Corporate social responsibility, customer satisfaction, and market value. *Journal of Marketing, 70*(4), 1–18. https://doi.org/10.1509/jmkg.70.4.001

Meher, B. K., Hawaldar, I. T., Mohapatra, L., Spulbar, C., & Birau, R. (2020). The Effects of Environment, Society and Governance Scores on Investment Returns and Stock Market Volatility. *International Journal of Energy Economics and Policy, 10*(4), 234–239. https://doi.org/10.32479/ijeep.9311

Nollet, J., Filis, G., & Mitrokostas, E. (2016). Corporate social responsibility and financial performance: A non-linear and disaggregated approach. *Economic Modelling, 52*, 400–407. https://doi.org/10.1016/j.econmod.2015.09.019

Przychodzen, J., & Przychodzen, W. (2013). Corporate Sustainability and Shareholder Wealth. *Journal of Environmental Planning and Management, 56*, 474–493. https://doi.org/10.1080/09640568.2012.685927

Zhang, J., Djajadikerta, H. G., & Zhang, Z. (2018). Does sustainability engagement affect stock return volatility? Evidence from the Chinese financial market. *Sustainability, 10*(12), 4417. https://doi.org/10.3390/su10103361

Chapter 6
ESG Data and Metrics

Marisa Laokulrach

Abstract In today's business landscape, environmental, social, and governance (ESG) considerations have become increasingly vital. ESG policies not only assist companies in addressing global challenges like climate change, resource scarcity, and social inequality but also empower those with robust ESG initiatives to minimize operational risks, streamline business costs, and uphold compliance with regulatory standards. This positions them as safer and more appealing investment choices.

ESG data plays a pivotal role for investors, companies, and other stakeholders keen on evaluating investments or businesses' sustainability and ethical impact. Numerous sources offer ESG data, each with varying degrees of reliability. This chapter explores ESG metrics and their dependability, ESG performance indicators, and the incorporation of ESG data in investment analysis and showcases examples of prominent ESG sources, ratings, and indices.

6.1 ESG Metric and Reliability

ESG metrics are the measurements of the effectiveness of a company's ESG policies. They provide valuable information about ESG risks and opportunities of a company, helping them to operate more ethically. ESG metrics are also useful as they facilitate industry benchmarking, which enables investors and other stakeholders to compare the ESG performance of a company against others in the same industry. Investors increasingly care about how the companies they invest in impact the world around them, and so companies with strong ESG performance tend to be seen as more attractive—and less risky—investments.

M. Laokulrach (✉)
International College, National Institute of Development Administration, Bangkok, Thailand
e-mail: marisa.lao@nida.ac.th

M. Bednárová, K. Soratana (eds.), *Environmental, Social, and Governance (ESG) Investment and Reporting*, https://doi.org/10.1007/978-3-031-84235-1_6

There is no universally recognized, mandatory ESG framework that companies must use. Instead, there are a variety of different frameworks that a company can select from. These frameworks provide guidance for companies in the identification, measurement, and documentation of their ESG commitments. ESG metrics make the process more scientific by outlining how different ESG topics should be measured and tracked. They also help to benchmark performance.

6.1.1 Benefits of ESG Metric to Company

Companies complying with ESG in their operations benchmark and standards and are measured by ESG metrics have the following benefits:

1. Proof of Operational Commitment—ESG metrics accommodate a company to prove its commitment to environmental, social, and governance issues such as climate change, the environment, and human rights. ESG metrics provide the data to back up such company claims.
2. Progress Indicator—ESG performance can determine the progress of complying with ESG in the operation progress. ESG metrics can also flag areas for concern, or areas of risk, allowing the company to make improvements.
3. Transparency—Transparency is increasingly important in the business world and is demanded by the company's stakeholders. ESG reporting is one of the important indicators of a company's operational transparency and governance.
4. Investment Decision—Most global investors consider ESG information when making investment decisions because ESG information is financially material to investment performance. ESG metrics provide a more rigorous way to evaluate and compare companies.
5. Brand Reputation—ESG metrics provide credibility to a company's sustainability management and development. ESG metrics support the company's reputation, credibility, and trustworthiness.

6.1.2 Reliability of ESG Data Sources

The wide variety of ESG frameworks and standards that are available offers flexibility for companies who can select the framework that they believe works best for their business. However, it also makes the ESG landscape more complex and can make it harder for stakeholders and investors to compare the ESG credentials of different companies as different metrics may be used depending on what framework they selected.

The reliability and quality of ESG data are critical considerations for investors, companies, and other stakeholders seeking to incorporate sustainability factors into decision-making processes. Understanding the strengths and limitations of ESG data is crucial for making informed choices and promoting transparency. Here are key points related to the reliability and quality of ESG data:

A. Data Accuracy

Maintaining the accuracy of reported environmental, social, and governance (ESG) data is a crucial concern, given that companies may occasionally provide incomplete or inaccurate information, leading to misguided assessments. Ensuring accuracy is fundamental for meaningful analysis, decision-making, and fostering trust among stakeholders. The following considerations and challenges are pertinent to the accuracy of ESG data:

Self-Reporting and Verification:
Companies often self-report their ESG data, raising concerns about accuracy and reliability. Independent third-party verification or certification processes can enhance data accuracy by providing external validation.

Quality of Data Sources:
ESG data accuracy is closely tied to the quality of underlying sources. Companies may utilize various sources, such as internal records, surveys, and external databases. Assessing source reliability and ensuring data quality during collection is vital for maintaining accuracy.

Data Validation and Quality Assurance:
Robust validation and quality assurance processes within organizations help identify and rectify errors before reporting. Automated tools and manual checks can verify the accuracy of data points and calculations. Technology, including data analytics and automation tools, can enhance ESG data accuracy. However, careful oversight is necessary to prevent errors in algorithms or data processing. Continuous monitoring and improvement contribute to data.

Consistency in Reporting:
Inconsistencies in defining and reporting ESG metrics can lead to inaccuracies. Standardized frameworks, like those from rating organizations, promote consistency. Companies should align their reporting practices with industry standards and communicate any changes.

External Factors and Events:
External events like natural disasters or regulatory changes can impact data accuracy. Companies should promptly update data in response to such events and conduct scenario analysis to assess external factors' effects on ESG metrics.

Technology and Automation:
Leveraging technology, including data analytics and automation tools, can enhance the accuracy of ESG data. However, reliance on technology also requires careful oversight to prevent errors in algorithms or data processing. Continuous monitoring and improvement of automated processes contribute to data accuracy.

B. Data Completeness

Data completeness is a vital element in ESG reporting, ensuring the presentation of a comprehensive and holistic picture of a company's sustainability performance. Incomplete data can result in misinterpretations and hinder stakeholders' ability to make well-informed decisions. Here are key considerations related to data completeness in ESG reporting:

Consistency and Comparability:
Maintaining consistency in measuring and reporting ESG metrics over time is essential for effective trend analysis. Inconsistencies can impede accurate assessments of a company's sustainability progress. Standardization initiatives, including the creation of industry-specific metrics, aim to improve comparability among companies in the same sector.

Scope and Coverage:
Companies must delineate the scope of their ESG reporting, clearly defining the operational boundaries and entities included in the report. Ensuring coverage across relevant business units, subsidiaries, and geographical locations is indispensable for presenting a complete representation of the company's impact.

Comprehensive Metrics:
ESG reporting should encompass a broad array of metrics within each of the three dimensions—environmental, social, and governance. This involves incorporating specific indicators relevant to the industry and addressing material issues for the company. The identification and reporting of key performance indicators (KPIs) are pivotal to ensuring that the most critical aspects of sustainability are not overlooked.

Stakeholder Engagement:
Actively engaging with stakeholders assists companies in identifying the ESG issues that hold the utmost significance to their audiences. This process aids in determining which data points are most relevant and should be included in reporting. Regular communication with stakeholders contributes to continuous improvement in data completeness.

Industry-Specific Metrics:
Recognizing the diverse ESG challenges faced by different industries, companies should include industry-specific metrics in their reporting. Standardization efforts by organizations like the Sustainability Accounting Standards Board (SASB) guide companies in determining the most relevant metrics for their sector.

Data Collection Processes:
The implementation of robust data collection processes is critical for ensuring completeness. This involves gathering data from internal systems, external sources, and stakeholders. Automation tools and technologies play a key role in streamlining data collection processes, reducing the risk of oversight, and ensuring more comprehensive coverage.

Integration with Business Processes:

Integrating ESG considerations into core business processes is crucial to ensuring that relevant data is consistently captured. Companies that embed sustainability into their day-to-day operations are more likely to have comprehensive and reliable ESG data.

External Verification and Assurance:

Engaging external parties, such as auditors or verification services, to review and validate ESG data can enhance its credibility and completeness. Independent assurance provides additional confidence to stakeholders that the reported information is accurate and comprehensive.

C. Data Timeliness

Timely reporting of environmental, social, and governance (ESG) data is crucial for stakeholders who rely on current information to make informed decisions. Outdated information may not accurately reflect a company's current sustainability practices. Regular reporting cycles and the use of real-time data platforms contribute to the timeliness of ESG information. Key considerations for data timeliness in ESG reporting are as follows:

Reporting Cycles:

Establishing regular reporting cycles for ESG data ensures consistent updates. Common reporting periods, such as annual, semi-annual, or quarterly updates, should be clearly communicated to stakeholders to help them anticipate when new data can be expected.

Real-Time Data Platforms:

Leveraging technology and real-time data platforms allows companies to provide more up to date ESG information. These platforms facilitate continuous data collection and reporting, benefiting stakeholders who require the latest information for decision-making.

Event-Driven Updates:

Timely updates in response to significant events or changes within the company or its operating environment are essential. This includes updates related to mergers, acquisitions, regulatory changes, or major sustainability initiatives. Companies should have mechanisms to communicate material updates promptly.

Integration with Financial Reporting:

Integrating ESG reporting with financial reporting processes contributes to timely disclosures. This ensures that ESG information is updated alongside financial performance data. Aligning ESG reporting with established financial reporting timelines improves overall reporting efficiency.

Technology and Automation:

Utilizing technology, such as data automation tools, streamlines the data collection and reporting process, reducing the time lag between data collection and publication. Automation helps ensure that ESG data is continuously monitored and updated as new information becomes available.

Regulatory Requirements:

Compliance with regulatory requirements related to ESG reporting timelines is essential. Companies should be aware of deadlines set by regulatory bodies and ensure they meet these obligations. Proactive engagement with evolving regulatory standards helps companies stay ahead of reporting requirements.

Communication of Data Delays:

Transparent communication with stakeholders is crucial if there are delays in reporting or unexpected challenges in data collection. Providing explanations for delays and setting expectations for when updated data will be available helps maintain trust.

Benchmarking and Performance Tracking:

Timely ESG data allows for accurate benchmarking and tracking of performance over time. Stakeholders can assess a company's progress in meeting sustainability goals and objectives. Consistent and timely reporting facilitates the identification of trends and areas for improvement.

By prioritizing the timeliness of ESG data, companies can meet stakeholder expectations, enhance transparency, and provide relevant information for decision-making and benchmarking purposes. Continuous improvement in reporting processes and technology adoption contributes to achieving greater timeliness in ESG disclosures.

D. Data Transparency

Transparency in the disclosure of environmental, social, and governance (ESG) data is indispensable for stakeholders to comprehend how metrics are calculated and reported. The absence of transparency can breed skepticism and impede trust in the reported information. Clear communication regarding methodologies, data sources, and any changes in reporting practices is key to ensuring the transparency of ESG data. Here are key considerations for data transparency in ESG reporting:

Reporting Methodologies:

Clearly articulate the methodologies used to collect, calculate, and report ESG metrics. This involves providing details on data sources, measurement units, and any assumptions made. Transparency in reporting methodologies aids stakeholders in understanding the basis of the reported data and promotes consistency.

Standardization and Frameworks:

Adherence to recognized ESG reporting frameworks, such as the Global Reporting Initiative (GRI) or Sustainability Accounting Standards Board (SASB), is crucial. Following established standards enhances comparability and transparency. The company should indicate the standards or frameworks used and how the company aligns with them.

Disclosure of Limitations:

Acknowledge and communicate any limitations or challenges associated with the reported ESG data. This may include data gaps, uncertainties, or instances where certain metrics are difficult to measure accurately. Transparently communicating limitations helps stakeholders interpret the information with the appropriate context.

Frequency and Timing of Reporting:

Clearly communicate the frequency of ESG reporting (e.g., annual, quarterly) and adhere to established reporting timelines. Consistent reporting contributes to transparency and enables stakeholders to anticipate when updates will be available. If there are delays or changes in reporting schedules, provide timely explanations to stakeholders.

Integration with Financial Reporting:

Integrate ESG information with financial reporting where possible. Present the links between financial and nonfinancial performance, demonstrating how sustainability factors are considered in overall business strategies. Integrating ESG information with financial disclosures enhances the completeness and transparency of reporting.

External Assurance and Verification:

Engage external parties for assurance and verification of ESG data. Independent audits or verification processes contribute to the credibility and transparency of reported information. Clearly communicate the outcomes of external assurance processes, including any identified areas for improvement.

Interactive and Accessible Reporting:

Utilize interactive and accessible reporting formats, such as online platforms or data visualizations, to make ESG information more user-friendly and accessible to diverse stakeholders. Ensure that the reported information is presented clearly and understandably.

Transparent Communication of Progress:

Transparently communicate progress made toward achieving sustainability goals and objectives. Clearly articulate the company's ambitions, targets, and timelines. Regularly update stakeholders on the status of ongoing initiatives and any changes in the company's approach to sustainability.

By emphasizing these considerations, companies can enhance the transparency of their ESG reporting, fostering trust and confidence among stakeholders. Transparent communication contributes to the credibility of the reported information and supports the broader goals of sustainability and responsible business practices.

E. Data Materiality

Identifying and reporting on material environmental, social, and governance (ESG) issues specific to a company's industry and operations is critical. Materiality assessments play a key role in prioritizing relevant factors, ensuring that disclosed information aligns with the company's actual impact. Stakeholder engagement and consultation processes can further enhance the identification of material ESG issues. Here are key considerations related to the data materiality of ESG reporting:

Identification of Material Issues:
Clearly articulate the process used for identifying material ESG issues. This involves assessing the potential impact of various sustainability factors on the company's financial performance and stakeholder interests. Engage with stakeholders, including investors, customers, employees, and communities, to understand their perspectives on material issues.

Stakeholder Input and Consultation:
Incorporate stakeholder input into the materiality assessment process. Engage in meaningful consultations with key stakeholders to identify their concerns, expectations, and priorities related to sustainability. Clearly communicate how stakeholder feedback informs the determination of material issues.

Relevance to Business Strategy:
Companies should demonstrate how material ESG issues align with their overarching business strategy. It is crucial to articulate how addressing these issues not only aligns with the company's mission, vision, and strategic objectives but also contributes significantly to long-term value creation and risk mitigation.

Industry-Specific Considerations:
Acknowledge industry-specific challenges and considerations related to ESG factors. Materiality assessments must encompass the distinctive risks and opportunities inherent in the company's sector. To ensure a comprehensive evaluation, refer to industry-specific reporting standards, such as those outlined by the Sustainability Accounting Standards Board (SASB), for guidance on identifying and addressing material issues.

Regular Review and Updating:
Regularly assess and revise materiality evaluations to mirror shifts in the business landscape, industry dynamics, and stakeholder expectations. Ensure transparent communication of any updates made to the materiality assessment process and its outcomes. This proactive approach supports ongoing transparency and keeps stakeholders informed of the evolving significance assigned to various factors impacting the business.

Financial Impact Assessment:
Evaluate the potential financial consequences of significant ESG issues on the company. Consider both direct financial effects and indirect impacts on brand value, customer loyalty, employee satisfaction, and regulatory compliance. Clearly articulate the methodology employed to assess financial impact and communicate the assessment results.

Clear Communication in Reports:
Effectively communicate the outcomes of the materiality assessment in ESG reports. Provide a transparent explanation of the criteria utilized, the identified material issues, and their relevance to the company's operations. Utilize accessible language and visuals to enhance stakeholders' comprehension of materiality considerations.

Dynamic Nature of Materiality:
Acknowledge the dynamic nature of materiality. Recognize that ESG issues may evolve over time, and the company's materiality assessments should adapt accordingly. Engage in ongoing dialogue with stakeholders to stay informed about emerging issues and changing expectations. By addressing these considerations, companies can enhance the materiality of their ESG reporting, focusing on issues that are most relevant to their business and stakeholders. This approach contributes to more meaningful and impactful sustainability reporting.

F. Data Governance and Assurance

Establishing robust data governance frameworks within organizations is paramount for ensuring the reliability and quality of ESG data. These frameworks encompass internal controls, data validation processes, and adherence to reporting standards. External assurance provided by independent auditors further bolsters confidence in the accuracy and completeness of reported ESG information. Stakeholders, by addressing these considerations, actively contribute to the ongoing improvement of the reliability and quality of ESG data, fostering greater transparency and accountability in sustainable business practices.

Data governance and assurance play crucial roles in ensuring the reliability, accuracy, and transparency of ESG data. These processes are essential for building trust among stakeholders, meeting regulatory requirements, and supporting informed decision-making. Here are key considerations related to data governance and assurance of ESG reporting:

Data Quality Standards:
Organizations should define and strictly adhere to data quality standards when dealing with environmental, social, and governance (ESG) data. This involves establishing clear criteria encompassing accuracy, completeness, timeliness, and consistency. Accuracy is crucial, requiring the precise representation of the company's environmental impact, social initiatives, and governance practices in the reported ESG data. Completeness emphasizes the need to capture all pertinent ESG metrics without omission, ensuring a comprehensive reflection of the organization's sustainability efforts.

Data Integration with Overall Governance:
Integrate ESG data governance seamlessly into the broader corporate governance structure, ensuring a cohesive and interconnected approach. Aligning ESG data governance with broader risk management and compliance processes is imperative to create a unified strategy that addresses environmental, social, and governance considerations in a comprehensive manner.

Data Security and Privacy:
Implementing robust data security measures is crucial to safeguard the confidentiality and integrity of ESG data. Organizations must prioritize the implementation

of advanced security protocols, encryption mechanisms, and access controls to prevent unauthorized access and ensure the protection of sensitive information.

Data Training and Awareness:

To enhance the competency of employees engaged in ESG data collection and reporting, organizations should institute comprehensive training and awareness programs. These initiatives aim to equip employees with the necessary knowledge and skills to effectively collect, manage, and report ESG data. These programs should cover various aspects, including data collection methodologies, reporting standards, and the significance of ESG metrics in the broader context of sustainability.

Continuous Improvement:

To ensure the ongoing enhancement of data governance practices, organizations should establish structured processes for continuous improvement. This involves creating mechanisms to systematically assess and refine data governance frameworks over time. Regular reviews are imperative to evaluate the effectiveness of existing practices and identify areas for enhancement.

External Assurance and Verification Process:

To ensure the credibility and reliability of ESG data, organizations should actively engage external auditors or assurance providers for independent review and verification. It is essential to select assurance providers with specialized expertise in sustainability reporting and auditing. This expertise ensures that the verification process comprehensively evaluates the accuracy and adherence to reporting standards specific to ESG metrics.

Continuous Monitoring and Auditing:

To ensure the ongoing integrity and reliability of ESG data, organizations should implement continuous monitoring and auditing processes. This involves regular reviews of data governance practices to assess the effectiveness of internal controls. Through these systematic evaluations, organizations can identify areas for improvement and enhance data quality and reporting practices.

Regulatory Compliance:

Organizations should ensure that their ESG reporting and assurance practices align with regulatory requirements. It is crucial to stay informed about any changes in regulations related to ESG disclosure and compliance. This proactive approach helps organizations adapt their reporting practices to remain in compliance with evolving regulatory standards.

By implementing robust data governance practices and engaging in external assurance processes, companies can enhance the reliability and credibility of their ESG data. These practices contribute to the broader goals of transparency, accountability, and responsible corporate citizenship.

6.2 Key ESG Metrics and Performance Indicators

Depending on the selected framework, a broad array of indicators falls under ESG metrics. The subsequent examples are organized within the three pillars—environmental, social, and governance—offering clarity on frequently employed ESG metrics and their corresponding measurement criteria.

6.2.1 Examples of Environmental Indicators (E)

Greenhouse Gas Emissions

The emission of greenhouse gases is pivotal in global warming, and companies significantly contribute to annual emissions. It is imperative to gauge emissions related to a company's operations, products, services, and supply chain. Governments globally are increasingly mandating greenhouse gas reporting. Metrics often revolve around carbon dioxide (or equivalent) quantified in tons or kilograms.

Air Pollution

Beyond greenhouse gases, a company's activities can contribute to air pollution, posing risks to the environment and human health. Measurement involves evaluating particle matter per aerodynamic diameter.

Energy Consumption

Given the substantial reliance of many businesses on energy, often derived from fossil fuels, quantifying energy consumption (typically in kilowatts per hour—kWh) serves as a valuable tool for monitoring and reducing consumption.

Water Consumption

As water scarcity becomes a pressing concern due to climate change, various industries utilize significant amounts of water, making water usage a critical ESG metric.

Resource Usage

The Earth's resources are finite, and excessive use can lead to ecosystem destruction and species extinction. Measuring resource depletion and land use provides insights into a company's impact on natural resources.

6.2.2 *Examples of Social Indicators*

Living Wages

Metrics are employed to evaluate a company's average wage concerning the cost of living, ensuring fair compensation, and enabling employees to afford necessities.

Diversity

The importance of diversity, especially at the executive level, is underscored. Metrics examining diversity percentages across various organizational levels offer valuable insights into workplace inclusivity, reflecting the representation of diverse demographics, including gender, ethnicity, and age, particularly in leadership positions.

Health and Safety

Metrics focused on incident reports and health and safety policies aim to ensure that work does not compromise the well-being of employees. This involves assessing workplace safety records, incident reports, and the effective implementation of health and safety policies.

Human Rights

Given the fundamental importance of human rights in society, companies should actively uphold them. Metrics include the monitoring of reports on human rights violations and the presence of company policies dedicated to human rights. This encompasses the commitment to preventing human rights violations in both the company's operations and its supply chain.

Social Impact

The measurement of a company's social impact involves assessing the positive or negative effects of its operations on society. Considerations include factors such as access to education, healthcare, or affordable housing, providing insights into the company's contribution to societal well-being.

6.2.3 *Examples of Governance Indicators*

Board Composition and Independence

The examination of the board of directors involves assessing its composition, with a focus on the presence of independent directors and the implementation of measures to mitigate conflicts of interest.

Executive Compensation

Metrics evaluating executive compensation practices, particularly when significantly exceeding employee pay, address concerns related to wealth inequality. The assessment concentrates on the fairness and transparency of compensation, analyzing the relationship between executive pay and that of the broader workforce.

Ethics and Anti-corruption

Key metrics in this category revolve around the existence and adherence to anti-corruption and bribery policies. The evaluation assesses the effectiveness of policies addressing ethics, anti-corruption, and bribery within the organization.

Shareholder Rights

An analysis of shareholder rights includes an examination of the rights and protections afforded to shareholders. This involves assessing elements such as voting rights and the extent to which shareholders can influence key decisions.

Stakeholder Engagement

When evaluating stakeholder engagement, the focus lies on the company's efforts to engage and address the concerns of various stakeholders, including shareholders, employees, and communities.

Transparency and Disclosure

The analysis of transparency and disclosure involves assessing the clarity and completeness of both financial and nonfinancial disclosures. The goal is to ensure that relevant information is communicated transparently to stakeholders.

Cybersecurity and Data Privacy

Metrics in this area assess measures taken to protect sensitive information, ensure data privacy, and manage cybersecurity risks. The evaluation encompasses the effectiveness of safeguards in place to protect against potential data breaches and privacy violations.

6.3 Utilizing Basic ESG Data in Investment Analysis

In the 1960s, investors began seeking avenues to make a positive impact with their investments through practices such as boycotting specific stocks or industries, known as socially responsible investing. Over time, socially responsible investments evolved from being a sideline player to a crucial consideration, prompting companies and industries to adapt. The first environmental, social, and governance (ESG) index emerged in 1990 in response to the increasing interest in ESG issues. ESG asset investments nearly doubled from 2015 to 2022, and this trend continues to gain momentum. The substantial growth in ESG investing has compelled companies to enhance the transparency of their reporting, placing a greater emphasis on sustainability in their operations.

6.3.1 Three Primary Drivers of ESG Investment

Changing Global Landscape

New global challenges, including regulatory pressure and increased scrutiny related to climate risk, sustainability trends, and demographic shifts, have emerged. The adoption of international standards, such as the UN Principles for Responsible Investment (PRI) and the Sustainable Development Goals (SDGs), has established a common framework for evaluating ESG performance globally.

Risk Management

ESG factors are recognized as indicators of a company's long-term sustainability and performance. Integrating these factors into investment decisions aids in identifying and managing risks that may not be evident through traditional financial analysis.

Technological Advancements

The advent of new tools, such as natural language processing, artificial intelligence, blockchain, and the Internet of Things, has facilitated higher-quality data collection and analysis. This has fostered increased trust in the transparency and accuracy of a company's operational reports among rating agencies and investors.

Research findings indicate that a significant majority of investors consider ESG information when making investment decisions, deeming it financially material to investment performance. However, the materiality of ESG information may systematically vary among countries, industries, and company strategies. For instance, different countries may prioritize issues like water pollution or corruption, while industries may be affected by climate change or human rights violations in the supply chain. The utilization of ESG data in investment analysis has become increasingly prevalent due to client demand and as part of the product development process. This has led to the creation of innovative financial instruments, such as green bonds, where proceeds are allocated to projects that enhance environmental outcomes. Understanding the structure and pricing of these contracts provides insights into investor preferences and the societal impact of such financial instruments.

6.3.2 *Key Ways of Utilizing ESG Data in Investment Analysis*

The integration of environmental, social, and governance (ESG) data into investment analysis has gained significance as investors aim to incorporate sustainability factors into their decision-making processes. ESG considerations offer insights beyond traditional financial metrics, enabling investors to assess risks, opportunities, and the overall sustainability of their investments. The following outlines key-ways in which ESG data is utilized in investment analysis:

Risk Assessment

- *Evaluate ESG-related Risks*: Assess a company's exposure to environmental, social, and governance (ESG) risks, including climate change, regulatory compliance, social controversies, and governance issues.

- *Environmental Risks*: Examine a company's vulnerability to environmental risks such as climate change, resource scarcity, and regulatory compliance.
- *Social Risks*: Analyze social risks, including labor practices, employee relations, and community impact.
- *Governance Risks*: Scrutinize governance structures and practices to identify potential risks related to board effectiveness, executive compensation, and transparency.

Opportunity Identification

- *Innovation and Efficiency*: Identify companies showcasing innovation and efficiency in managing environmental resources, reducing waste, and adopting sustainable practices.
- *Social Impact*: Evaluate investments that positively contribute to social development, community engagement, and diversity and inclusion.
- *Governance Practices*: Recognize companies with robust governance practices promoting ethical behavior, stakeholder engagement, and long-term value creation.

Performance Metrics

- *Long-Term Financial Performance*: Explore the positive correlation between strong ESG performance and long-term financial outperformance. Use ESG metrics as indicators of a company's ability to manage risks and opportunities over the long term.
- *Alpha Generation*: Integrate ESG factors into alpha generation strategies to gain a competitive advantage by identifying companies likely to outperform due to strong ESG practices.

Regulatory Compliance and Legal Risks

- *Compliance Assessment*: Evaluate a company's compliance with existing and emerging ESG-related regulations affecting future business operations. Assess compliance with environmental regulations, labor laws, and other relevant regulations.
- *Legal and Regulatory Risks*: Assess potential legal and regulatory risks associated with ESG issues, such as lawsuits related to environmental violations or labor practices.

Integration into Investment Models

- *ESG Integration*: Integrate ESG factors into traditional financial models for a more comprehensive investment analysis. Assess the impact of ESG factors on a company's valuation, revenue growth, and cost structure.
- *Scenario Analysis*: Conduct scenario analysis to evaluate how different ESG scenarios, such as climate-related events, may impact investment portfolios.

Disclosure and Transparency

- *Transparent Reporting*: Favor investments in companies providing transparent and comprehensive ESG disclosures. Transparent reporting enhances the ability to assess the materiality and relevance of ESG issues.

ESG Funds and Index Investing

- *ESG-focused Funds*: Consider investing in funds or portfolios explicitly following ESG principles.
- *Benchmarking against ESG Indices*: Evaluate investment performance against ESG indices to ensure alignment with sustainability goals.

As ESG considerations gain prominence, investors increasingly recognize the value of integrating ESG data into their decision-making processes. The objective is to achieve a holistic understanding of a company's performance, considering not only financial metrics but also the broader impact on the environment, society, and governance practices.

6.4 Leading ESG Rating and Index

6.4.1 Leading ESG Rating Organization

Numerous organizations play a pivotal role in providing environmental, social, and governance (ESG) ratings and assessments, aiding investors and stakeholders in evaluating the sustainability and ethical practices of businesses. Diverse methodologies and criteria are employed to assess a company's performance in crucial ESG areas. One of the notable ESG rating organizations includes World Economic Forum (WEF). Other important initiatives that provide ESG reporting standards and frameworks are discussed in Chap. 7.

World Economic Forum (WEF) (https://www.weforum.org)

The World Economic Forum (WEF) stands as a distinguished international non-profit organization, established in 1971. Its primary mission is to convene leaders from diverse sectors worldwide, fostering collaboration to enhance the state of the world. Serving as a platform for dialogue, idea exchange, and cooperative efforts, the WEF addresses global challenges through the following key objectives:

The WEF underscores the significance of public-private cooperation as a pivotal approach to tackling global challenges. Encouraging collaboration among governments, businesses, and civil society, the organization aims to inspire innovative solutions. The WEF directs its efforts toward addressing critical global issues such as climate change, economic inequality, technological advancements, healthcare, and education. Its overarching goal is to instigate positive change in these vital areas.

In 2020, the WEF introduced the concept of "Stakeholder Capitalism." This transformative initiative advocates for businesses to prioritize the interests of all stakeholders, including customers, employees, communities, and the environment, transcending the traditional focus solely on shareholders. Aligned with the Stakeholder Capitalism initiative, the WEF unveiled universal performance metrics in 2020. These metrics are meticulously designed to be universally applicable, irrespective of industry or business model, with the aim of promoting sustainable and responsible business practices.

The WEF contributes significantly to the global discourse by producing research reports, white papers, and publications. Covering a spectrum of topics related to the global economy, technology, sustainability, and societal issues, these resources play a vital role in informed decision-making and policy discussions. Amid the unprecedented challenges posed by the COVID-19 pandemic, the WEF actively facilitated discussions on global responses to the health crisis and its profound economic and social impacts.

6.4.2 Leading ESG Indices and Scores

Environmental, social, and governance (ESG) indices are instrumental in incorporating ESG criteria into the selection and weighting of components. These indices serve as benchmarks for sustainable and socially responsible investment strategies. Key ESG indices include:

MSCI ESG Indexes (https://www.msci.com)

The MSCI ESG Indexes, curated by MSCI Inc., represent a globally acknowledged suite of environmental, social, and governance (ESG) indices meticulously designed to assess and benchmark the ESG performance of companies. As a premier provider of investment decision support tools and services, MSCI's ESG Indexes hold a pivotal role in shaping sustainable and socially responsible investment strategies. The MSCI ESG Indexes stand as a comprehensive family encompassing various regions, countries, and sectors. Tailored to incorporate ESG criteria into the selection and weighting of components, these indices serve as benchmarks for investors aiming to align their portfolios with sustainable and responsible investment principles.

MSCI extends a diverse array of ESG indices with expansive global coverage. Noteworthy indices within this spectrum include the MSCI World ESG Index, MSCI Emerging Markets ESG Index, and others. These indices traverse different regions and markets, empowering investors to evaluate the ESG performance of companies on a global scale.

Integral to the methodology of constructing MSCI ESG Indexes is the integration of MSCI ESG Ratings and data. These ratings furnish a quantitative evaluation of a company's ESG performance, encompassing a broad spectrum of factors within the environmental, social, and governance dimensions. Ratings are assigned on a scale from AAA to CCC, with AAA indicative of superior ESG performance.

MSCI ESG Ratings involve key ESG issues pertinent to each industry. These issues, diverse across sectors, encompass considerations such as carbon emissions, labor practices, corporate governance, and community relations. The objective is to offer a nuanced evaluation aligned with the distinct challenges and opportunities within each industry.

Beyond positive ESG performance, MSCI ESG Ratings incorporate considerations of ESG controversies associated with companies. This element aids investors in identifying companies embroiled in disputes related to environmental damage, labor issues, ethical concerns, or other ESG-related matters. MSCI ESG Indexes have gained widespread adoption among investors seeking to infuse ESG considerations into their investment strategies. These indices facilitate the alignment of portfolios with sustainability goals, the mitigation of risks linked to poor ESG performance, and the potential contribution to positive social and environmental impact.

Dow Jones Sustainability Indices (DJSI) (https://www.spglobal.com)

The Dow Jones Sustainability Indices (DJSI) hold a distinguished position as globally recognized benchmarks meticulously designed to evaluate the sustainability performance of companies across diverse industries. Published by S&P Dow Jones Indices, these indices have become pivotal in the assessment of environmental, social, and governance (ESG) factors, exerting a profound influence on investment decision-making. This comprehensive exploration delves into the key features and significance of the Dow Jones Sustainability Indices, elucidating their role in shaping sustainable investing practices.

At its core, the Dow Jones Sustainability Indices serve as benchmarks within financial markets, offering a systematic approach to evaluating the sustainability endeavors of companies. By incorporating a range of ESG criteria, these indices furnish investors with a valuable tool to integrate sustainability considerations into their investment strategies, transforming sustainability from a qualitative concept into a measurable and comparable metric.

The DJSI comprises a diverse array of indices tailored to different segments of the market. Among the prominent ones are the DJSI World, DJSI North America, and DJSI Europe. These regional indices enable investors to evaluate sustainability performance on a global scale and within specific geographical contexts, recognizing the contextual nuances that may affect companies' ESG practices. Companies included in the Dow Jones Sustainability Indices undergo a rigorous evaluation process based on a comprehensive set of ESG criteria. This evaluation encompasses environmental considerations, such as climate strategy and water-related risks; social factors, including labor practices and corporate citizenship; and governance aspects, such as board structure and executive compensation. This holistic approach ensures a thorough examination of a company's sustainability performance across key dimensions.

Acknowledging the distinct sustainability challenges faced by different industries, the DJSI incorporates industry-specific criteria into its assessment methodology. This tailored approach ensures that companies are evaluated in a nuanced manner, considering the unique ESG factors relevant to each sector. By recognizing industry-specific challenges, the indices provide a more accurate reflection of a company's sustainability performance.

Inclusion in the Dow Jones Sustainability Indices is a testament to a company's commitment to sustainability. The selection process is highly competitive, with only companies demonstrating leadership and robust performance across various ESG dimensions earning a coveted place in these prestigious indices. This exclusivity enhances the credibility and reliability of the indices as indicators of sustainable business practices.

To maintain the relevance and accuracy of sustainability assessments, the Dow Jones Sustainability Indices undergo an annual review. This iterative process allows the indices to reflect changes in corporate sustainability performance over time, ensuring that investors have access to up-to-date information. The commitment to regular reviews underscores the dynamic nature of sustainability and the need for continuous monitoring to capture evolving ESG trends.

Refinitiv ESG Scores (https://www.refinitiv.com/fr/sustainable-finance/esg-scores)

As part of the London Stock Exchange Group (LSEG), Refinitiv provides ESG scores and data to help investors integrate sustainability considerations into decision-making processes. These scores evaluate a company's performance in key ESG areas, contributing to a holistic investment analysis.

Refinitiv ESG (environmental, social, and governance) Scores are instrumental in offering a comprehensive evaluation of a company's commitment to sustainability and ethical practices. Developed by Refinitiv, a prominent global provider of financial market data and infrastructure, these scores have gained widespread usage among investors, analysts, and corporations. The scores serve as a valuable tool to assess a company's performance across key dimensions of sustainability and responsible business practices. Here are key components of Refinitiv ESG Scores:

Environmental Criteria

This metric assesses a company's commitment to reducing its carbon footprint and adeptly managing risks associated with climate change. Additionally, it scrutinizes the effective utilization of natural resources, covering aspects like water and energy management. This criterion goes beyond and evaluates a company's impact on ecosystems and biodiversity, recognizing its pivotal role in contributing to environmental conservation efforts.

Social Criteria

This criterion involves an in-depth analysis of how a company treats its employees, encompassing crucial aspects such as labor rights, diversity, and the overall well-being of the workforce. It includes indicators that assess a company's engagement and influence on the communities in which it operates, providing a measure of the organization's social responsibility. Moreover, this criterion examines how companies navigate and address the social impacts associated with their products or services.

Governance Criteria

This criterion assesses the efficiency and transparency of a company's governance structure, guaranteeing strong oversight and ethical decision-making. It evaluates a company's dedication to ethical business practices and its initiatives to prevent corruption within its operations. By scrutinizing the rights and treatment of shareholders, this criterion ensures the implementation of fair and equitable practices in corporate governance.

Refinitiv ESG Scores draw from a diverse range of sources, including regulatory filings, company reports, and news articles. This comprehensive approach ensures a well-rounded assessment of a company's sustainability performance. Companies are scored on a scale, allowing for relative comparisons within industries or sectors. This scoring system provides a nuanced evaluation of companies' performance in the context of their specific industry or sector. The scores consider the materiality of different ESG factors, considering industry-specific considerations. This ensures that companies are assessed on factors relevant to their business context. ESG Scores undergo regular updates to reflect changes in a company's performance. This iterative process ensures that investors have access to timely and relevant information for informed decision-making.

Investors leverage Refinitiv ESG Scores to inform their investment decisions, integrating sustainability factors into their portfolios. The scores serve as a crucial guide for socially responsible investing. Corporations and investors utilize ESG Scores to identify and manage potential environmental, social, and governance risks. This proactive approach contributes to the mitigation of risks associated with unsustainable practices.

Example of ESG Rating Methodology: Using MSCI ESG Ratings

MSCI ESG Ratings are industry-relative measures and are determined at the company level. They are assessed on a global seven-band scale, ranging from AAA (the highest ESG Rating) to CCC (the lowest ESG Rating). The evaluation process considers two to seven Environmental and Social Key Issues, selected based on a company's exposure to potential ESG risks influenced by industry and market dynamics.

Each company is subjected to an evaluation on the Governance Pillar, which comprises six Key Issues related to Corporate Governance and Corporate Behavior Themes. This evaluation measures the gap between a company's governance practices and the best practices in the industry.

When applicable, the ESG Rating takes into account a company's alignment with market demands for products and services that make a positive contribution to the environment or society. MSCI ESG Ratings also incorporate a company's management measures in addressing overall ESG risks and opportunities. These measures are typically assessed through governance structures, policies, targets, quantitative performance metrics, and consideration of relevant controversies. The hierarchical representation of MSCI ESG ratings can be found in Fig. 6.1.

The top-level assessment is the overall Company ESG Rating, an industry-relative seven-point letter rating scale from AAA to CCC. These assessments are not absolute but are explicitly intended to be interpreted relative to a company's industry peers. The Company ESG Rating is derived from the final Industry-Adjusted Company Score, based on an assessment of the underlying data available at the last ESG Rating action date as shown in Fig. 6.2. An example of leading companies in different countries with MSCI assessment is presented in Table 6.1.

6.5 Summary

The reliability of environmental, social, and governance (ESG) data sources hinges on various factors such as accuracy, completeness, timeliness, transparency, materiality, and governance. Challenges in data accuracy arise from issues like self-reporting, the quality of data sources, and the impact of external factors. Emphasizing standardization, stakeholder engagement, industry-specific metrics, and integration

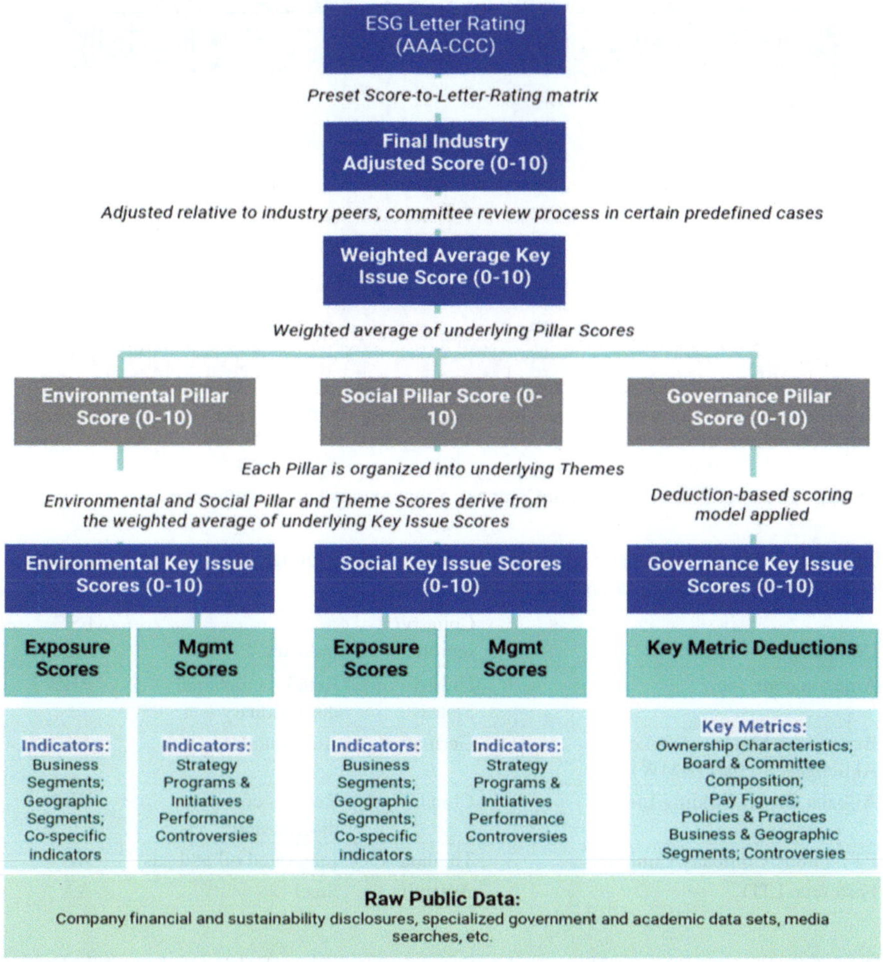

Fig. 6.1 Hierarchy of MSCI ESG ratings (www.msci.com/documents/1296102/34424357/MSCI+ESG+Ratings+Methodology.pdf)

with business processes is crucial. Transparency in reporting methodologies, adherence to standards, disclosure of limitations, and interactive reporting formats are vital for ensuring data transparency. Establishing robust data governance frameworks, adherence to quality standards, integration with overall governance, data security, training, continuous improvement, external assurance, and regulatory compliance are essential for ensuring ESG data's reliability and credibility.

ESG metrics and performance indicators are categorized into three pillars: environmental, social, and governance. Environmental indicators include greenhouse gas emissions, air pollution, energy consumption, water consumption, and resource usage. Social indicators encompass living wages, diversity, health and safety, human rights, and social impact. Governance indicators include board composition,

Letter Rating	Leader/Laggard	Final Industry-Adjusted Company Score
AAA	Leader	8.571* - 10.0
AA	Leader	7.143 - 8.571
A	Average	5.714 - 7.143
BBB	Average	4.286 - 5.714
BB	Average	2.857 - 4.286
B	Laggard	1.429 - 2.857
CCC	Laggard	0.0 - 1.429

**Appearance of overlap in the score ranges is due to rounding error. The 0-to-10 scale is divided into seven equal parts, each corresponding to a letter rating.*

Fig. 6.2 The final Industry-Adjusted Company Score mapped to a letter rating

Table 6.1 Examples of leading companies with MSCI ESG ratings (https://www.msci.com/our-solutions/esg-investing/esg-ratings-climate-search-tool)

Company	Country/ region	Industry	ESG ratings
Amazon.com, Inc.	United States	Retail—consumer discretionary	BBB
Bayerische Motoren Werke Aktiengesellschaft (BMW)	Germany	Automobiles	AA
Alibaba Group Holding Limited	China	Retail—consumer discretionary	BBB
PTT Public Company Limited	Thailand	Integrated oil and gas	AA
Fortescue LTD	Australia	Steel	AA

executive compensation, ethics and anti-corruption, shareholder rights, stakeholder engagement, transparency and disclosure, and cybersecurity and data privacy.

Investors increasingly consider ESG factors as indicators of long-term sustainability and performance, aided by advancements in data collection technologies. The materiality of ESG information varies across countries, industries, and company strategies. Keyways in which ESG data is utilized in investment analysis include risk assessment, opportunity identification, performance metrics, regulatory compliance evaluation, integration into investment models, disclosure and transparency considerations, and the rise of ESG-focused funds and index investing. ESG considerations provide investors with insights beyond traditional financial metrics, allowing for a comprehensive assessment of risks, opportunities, and the overall sustainability of investments. Numerous organizations, such as the Global Reporting Institute (GRI), Sustainability Accounting Standards Board (SASB), International Sustainability Standards Board (ISSB), and World Economic Forum (WEF), play a crucial role in evaluating sustainability and ethical practices. ESG indices, including MSCI ESG Indexes, Dow Jones Sustainability Indices (DJSI), NASDAQ CRD Global Sustainability Index, and Refinitiv ESG Scores.

6.6 Discussion Questions

1. How can companies ensure the reliability of their ESG data, considering factors like accuracy, completeness, and transparency?
2. In what ways can companies effectively engage with stakeholders to gather meaningful ESG data?
3. How do investors use ESG data in their decision-making processes, and what role does it play in assessing long-term sustainability?
4. How has technology, such as artificial intelligence and blockchain, improved the collection and analysis of ESG data?
5. What role do ESG metrics play in enhancing a company's reputation and brand value?
6. How do you foresee the future development of ESG metrics and data management?

References

Alsayegh, M. F., Abdul Rahman, R., & Homayoun, S. (2020). Corporate economic, environmental, and social sustainability performance transformation through ESG disclosure. *Sustainability, 12*(9), 3910. https://doi.org/10.3390/su12093910

Amel-Zadeh, A. (2018). *Social responsibility in capital markets: A review and framework of theory and empirical evidence*. Working paper, Saïd Business School. https://papers.ssrn.com/sol3/Papers.cfm?abstract_id=2664547

Asvathitanont, C., & Tangjitprom, N. (2020). The performance of environmental, social, and governance Investment in Thailand. *International Journal of Financial Research, 11*(6), 253. https://doi.org/10.5430/ijfr.v11n6p253

Atan, R., Alam, M. M., Said, J., & Zamri, M. (2018). The impacts of environmental, social, and governance factors on firm performance. *Management of Environmental Quality: An International Journal, 29*(2), 182–194. https://doi.org/10.1108/meq-03-2017-0033

Bauer, R., Koedijk, K. G., & Otten, R. (2005). International evidence on ethical mutual fund performance and investment style. *Journal of Banking & Finance, 29*(7), 1751–1767.

Bauer, R., Derwall, J., & Otten, R. (2007). The ethical mutual fund performance debate: New evidence from Canada. *Journal of Business Ethics, 70*(2), 111–124.

Becker, G. (1971). *The economics of discrimination* (2nd ed.). Chicago University Press.

Bénabou, R., & Tirole, J. (2010). Individual and corporate social responsibility. *Economica, 77*(305), 1–19.

Brammer, S., Brooks, C., & Pavelin, S. (2006). Corporate social performance and stock returns: UK evidence from disaggregate measures. *Financial Management, 35*(3), 97–116.

Brandon, R.G., & Krueger, P. (2018). *The sustainability footprint of institutional investors*. Swiss Finance Institute Research Paper No. 17-05 (January). https://papers.ssrn.com/sol3/papers.cfm?abstract_id=2918926

Camilleri, M. A. (2015). Environmental, social and governance disclosures in Europe. *Sustainability Accounting, Management and Policy Journal, 6*(2), 224–242. https://doi.org/10.1108/sampj-10-2014-0065

Cek, K., & Eyupoglu, S. (2020). Does environmental, social and governance performance influence economic performance? *Journal of Business Economics and Management, 21*(4), 1165–1184. https://doi.org/10.3846/jbem.2020.12725

Cheng, B., Ioannou, I., & Serafeim, G. (2014). Corporate social responsibility and access to finance. *Strategic Management Journal, 35*(1), 1–23.

Dhaliwal, D. S., Li, O. Z., Tsang, A., & Yang, Y. G. (2011). Voluntary nonfinancial disclosure and the cost of equity capital: The initiation of corporate social responsibility reporting. *Accounting Review, 86*(1), 59–100. https://doi.org/10.2308/accr.00000005

Dichev, I. D., Graham, J. R., Harvey, C. R., & Rajgopal, S. (2013). Earnings quality: Evidence from the field. *Journal of Accounting and Economics, 56*(2–3 Supplement 1), 1–33. https://doi.org/10.1016/j.jacceco.2013.05.004

Dimson, E., Karakaş, O., & Li, X. (2015). Active ownership. *Review of Financial Studies, 28*(12), 3225–3268.

Harymawan, I., Putra, F. K. G., Fianto, B. A., & Wan Ismail, W. A. (2021). Financially distressed firms: Environmental, social, and governance reporting in Indonesia. *Sustainability, 13*(18), 10156. https://doi.org/10.3390/su131810156

Plastun, A., Makarenko, I., Yelnikova, Y., & Makarenko, S. (2019). Environmental, social and governance investment standardization: Moving towards a sustainable economy. *Environmental Economics, 10*(1), 12–22. https://doi.org/10.21511/ee.10(1).2019.02

Ruan, L., & Liu, H. (2021). Environmental, social, governance activities and firm performance: Evidence from China. *Sustainability, 13*(2), 767. https://doi.org/10.3390/su13020767

Senadheera, S. S., Withana, P. A., Dissanayake, P. D., Sarkar, B., Chopra, S. S., Rhee, J. H., & Ok, Y. S. (2021). Scoring environment pillar in environmental, social, and governance (ESG) assessment. *Sustainable Environment, 7*(1), 1960097. https://doi.org/10.1080/27658511.2021.1960097

Tamimi, N., & Sebastianelli, R. (2017). Transparency among S&P 500 companies: An analysis of ESG disclosure scores. *Management Decision, 55*(8), 1660–1680. https://doi.org/10.1108/md-01-2017-0018

Tarmuji, I., Maelah, R., & Tarmuji, N. H. (2016). The impact of environmental, social and governance practices (ESG) on economic performance: Evidence from ESG score. *International Journal of Trade, Economics and Finance, 7*(3), 67–74. https://doi.org/10.18178/ijtef.2016.7.3.501

Yeoh, P. (2021). Sustainability of environmental, social and governance the sustainability of environmental, social and governance (ESG) reporting in the US and the UK. *Business Law Review, 42*(Issue 6), 272–281. https://doi.org/10.54648/bula2021038

Chapter 7
Impact Investing and the Effectiveness of Sustainable Finance

Marta Miranda García

Abstract The current landscape, marked by significant global challenges such as climate change, economic and social inequality, and the depletion of natural resources, demands effective responses and urgent measures, also from the realm of finance.

Thus, in this chapter, we will address how finance responds to this global call to action and adapts to the need for transitioning toward a sustainable economic and financial model. This involves providing concrete solutions and committed actions, giving rise to what is commonly known as sustainable finance.

The objective of this chapter is to comprehend the essential role of financial decisions in achieving sustainability, as well as the main tools or instruments used in the financial sector to contribute to the transition. To achieve this, various central themes will be developed, organized into subsections, allowing students to discover why sustainable finance is an essential driving force for change. These subsections include the following: context and sustainable finance; fundamentals of sustainable finance; sustainable finance products and instruments, and evaluating the effectiveness of impact measurement methodologies.

Finally, a brief recapitulation of the key points of this topic will be conducted, along with the main conclusions, prompting reflection on the impact and future of sustainable finance. Emphasis will be placed on the need to promote greater adoption of sustainable financial practices, with financial education playing an indispensable role.

M. M. García (✉)
Department of Financial Economics and Accounting, University Pablo de Olavide, Seville, Spain
e-mail: immirgar@upo.es

M. Bednárová, K. Soratana (eds.), *Environmental, Social, and Governance (ESG) Investment and Reporting*, https://doi.org/10.1007/978-3-031-84235-1_7

7.1 Context and Sustainable Finance

Sustainable finance has evolved within a context marked by unstoppable global trends, as illustrated in Table 7.1, which have driven its relevance in recent decades. These trends reflect the growing need to consider not only economic but also social and environmental factors in financial decision-making.

These circumstances, without which sustainable finance and its evolution cannot be properly understood, exhibit certain characteristics: they are critical, global, cross-cutting, and interdependent situations. In summary, the context that explains what sustainable finance is and its evolution is outlined below.

As we will see below, in this scenario, sustainable finance plays a crucial role in the effective response to these critical situations and new risks in the global economy.

Table 7.1 Trends affecting the financial system

Trend	Description
Climate change and environmental crises	The concern for climate change and environmental degradation is growing, driving the demand for sustainable financial solutions
Social awareness and corporate responsibility	Increased awareness of the social and environmental impacts of companies is leading to pressure for more ethical and sustainable practices
Regulation and government policies	Governments and regulators are implementing policies and regulations that promote sustainability
Conscious investors and consumers	Investors and consumers actively seek investments and products that reflect their sustainability values
Evolved financial markets	Markets adapt to accommodate sustainable financial instruments and create sustainable investment indices
Technology and data	Technological advances enable greater transparency and traceability in sustainable investments, as well as the collection and analysis of sustainability data
United Nations Sustainable Development Goals (SDGs)	The Sustainable Development Goals (SDGs) provide a global framework for addressing social and environmental challenges and guide investment strategies
Economic crises and pandemics	Economic crises and pandemics have underscored the need to consider sustainability in financial decisions
Perception of high uncertainty	Managing uncertainty has become a significant challenge for governments, businesses, and individuals today
Financialization	The growing role of financial interests and activities in the functioning of national and international economies

7.1.1 The Broad Concept of Sustainable Finance

Sustainable finance (hereafter SF) has garnered significant attention in recent years, especially since 2015 with the United Nations' global call for sustainable development. Business schools, financial institutions, financial markets, and regulators, as well as individual and institutional investors, have all focused on this topic. Despite its increasing importance, defining SF remains challenging (Cunha et al., 2021).

The concept of sustainable finance is not universal, causing confusion and ambiguity due to its multidimensional and complex nature.

From a practitioner's perspective, SF presents itself in different dimensions, capstones, or themes. Viewed as a system, sustainable finance comprises interrelated dimensions, including financial assets, financial markets, business sustainability, the planet, norms, and framework.

A broad perspective defines SF as the intersection between finance and the United Nations' 17 Sustainable Development Goals (SDGs). In this view, SF encompasses all economic resources and instruments necessary for the transition and achievement of the SDGs, efficiently channeling resources toward sustainable activities. The underlying concept is that the sustainability transition requires substantial investment, with the United Nations estimating a range of $5 trillion to $7 trillion to achieve the SDGs (Kumar et al., 2022).

On the other hand, a more specific definition is used by certain international bodies such as the European Commission or the World Bank, referring to SF as the process of integrating environmental, social, and governance (ESG) considerations into investment decisions in the financial sector. This leads to more long-term investments in sustainable economic activities and projects.

From the perspective of business management, a definition that considers the objectives of all stakeholders refers to SF as everything related to the study of financial decision-making, investment decisions, and financing aligned with various sustainability criteria. Therefore, sustainable finance incorporates sustainability as an integral part of the value proposition and value creation logic, striving for triple-bottom-line performance measurement.

In essence, SF is complex because it addresses a complex problem. The root of this complexity lies in the fact that the concept can pose significant challenges to conventional finance and/or paradoxes (Hahn et al., 2018; Yan et al., 2019).

One such challenge is the ambiguous relationship between financial logic and sustainability goals. Organizations pursuing sustainability objectives face inherent tensions between interdependent and conflicting goals when attempting to combine (or optimize) the pursuit of maximum profitability with the need to cease certain unsustainable activities.

This means that financial logic can play both a facilitating and restrictive role in implementing sustainable finance, depending on the prevalence of financial logic in society. Only when there is a moderate prevalence of financial logic can actors deviate from ultimate financial goals and still access these means, allowing for the reconciliation of these objectives.

According to scholars, sustainable finance can contribute to the development of a more comprehensive finance theory, providing a starting point for analyzing projects, portfolio management, business assessment, shareholder interaction, and public sector policy analysis that encompasses the social and ecological nature of finance (Aracil & Sancak, 2023).

Another significant paradox is the notion of "financialization," which has been identified as one of the reasons contributing to social and environmental problems. Thus, it raises questions about whether finance can genuinely contribute to sustainable development, especially considering that financial gains have been extracted even from the current climate disaster.

The confusion about what SF is has also manifested in its evolution over time with SF being identified with its subthemes. Some terms are sometimes used interchangeably, such as green finance, impact investments, etc.

SF was initially primarily associated with so-called socially responsible investments. However, between 2015 and 2020, in response to the Paris Agreement and the launch of the Sustainable Development Goals (SDGs), the focus shifted to impact investments and innovative financial instruments.

Coupled with this dynamic concept of SF, there has also been a profound transformation of SF, evolving from a niche market to a substantial section of the financial industry. This transformation is marked by increased participation from investors and regulators in the field (Ahlström & Monciardini, 2021).

7.1.2 The Contribution of Sustainable Finance to Sustainable Development

An essential aspect is how sustainable finance (SF) can contribute to sustainable development. Sustainable development is based on creating long-term value, and to achieve this, The 2030 Agenda for Sustainable Development was established, containing 17 Sustainable Development Goals (SDGs) that are integrated and interconnected (United Nations, 2015).

The SDGs generally encompass actions and policies aimed at mitigating the negative externalities of human activity, focusing on social inclusiveness (SDGs 1–11, 15, 16), ecological inclusiveness (SDGs: 1, 2, –9, 11–15), and relational inclusiveness (SDGs: 3, 4, 10–14, 16, 17) (Gupta & Vegelin, 2016). Interactions occur among SDGs, which can be both positive and negative, depending on key factors such as geographical context, resource endowments, time horizon, and governance (Nilsson et al., 2018).

To achieve the SDGs, financing is necessary, requiring substantial financial resources (it is estimated that an investment of at least USD 4–6 trillion per year will be required for a global transformation to a low-carbon economy in line with Paris Agreement objectives). These funds should come not only from state contributions but also, and especially, from companies and investors.

However, the shortage of funds is not the greatest problem in achieving the SDGs (according to the United Nations Development Programme, financing the SDGs would represent only 1% of global wealth). The challenge lies in directing financial resources toward the world's economies that are more vulnerable and in need, as well as toward areas of the economy essential for sustainability.

SF, as defined, serves as the facilitating and accelerating vehicle for this to occur and can play an instrumental role in the transition to a more sustainable world where both current and future needs are met.

SF can fulfill this crucial role by reformulating the international public and private financial architecture, integrating sustainability values and their impact into the financial decision-making process. This involves considering the creation of long-term value for the financial system, the environment, and society, as well as adapting the financial system to the sustainable development goals and correcting inadequate pricing.

From the perspective of public finances, it is not only necessary to increase financial contributions to sustainability (SDG financing needs have increased by 56% from $2.5 trillion to $3.9 trillion) but also to establish national and global policy frameworks that strengthen the driving forces of the real economy and create favorable conditions for sustainable investments. This could involve funding long-term innovation and societal infrastructure needs and accelerating the transition to a low-carbon economy and efficient resource use.

Moreover, public finances play an essential role in strengthening cooperation between the main market participants and regulatory entities in the financial sector to enhance financial stability, transparency, trust, and asset pricing. This includes improving the assessment and management of long-term risk and creating intangible value, including environmental, social, and governance aspects.

The United Nations Secretary-General's SDG Stimulus to Deliver Agenda 2030 also sets objectives to stimulate sustainable development, such as tackling the high cost of debt and rising risks of debt distress. This includes converting short-term high-interest borrowing into long-term (more than 30 years) debt at lower interest rates. Additionally, there is a call to massively scale up affordable long-term financing for development, especially by strengthening the capital base of multilateral development banks (MDBs), improving the terms of their lending, and aligning all financing flows with the SDGs. Furthermore, contingency financing to countries in need, including integrating disaster and pandemic clauses into all sovereign lending, is proposed, along with more automatic issuance of Special Drawing Rights (SDRs) in times of crisis.

From the perspective of private finance, companies, and/or investors, according to Schoenmaker and Schramade (2019), sustainable finance (SF) facilitates businesses and investors in making financial decisions directed toward sustainable practices by redefining their objectives. This involves adding social and environmental dimensions to the achievement of economic-financial results with a long-term vision.

SF can play a leading role in allocating investment to sustainable corporations and projects, thereby accelerating the transition. When allocating funds, finance can help make strategic decisions about sustainable goals. In this way, investors can influence the companies in which they invest and drive them toward more sustainable practices. SF also assesses pricing risk and can help solve uncertainties peculiar to environmental questions.

In essence, SF allows the incorporation of social and environmental aspects that correlate with sustainable development pillars into general finance theory (Fullwiler, 2016); fundamental pillar of sustainable development, especially social and environmental, relates to specific aspects of sustainable finance and, therefore, refers to the financial decisions involved in achieving those sustainable development goals.

7.2 Fundamentals of Sustainable Finance

We have already mentioned that the key element of sustainable finance (SF) lies in reformulating financial logic by incorporating not only risk-adjusted returns in financial decision-making but also the social and environmental impact in the long term.

Achieving sustainability requires a financially sustainable model as well because economic activities or investments can only be profitable if they are sustainable. Otherwise, in the future, we will all be "literally" dead (an adaptation of the famous phrase by M. Keynes).

Thus, the sustainability principle has strong implications for intertemporal preferences, for a given level of technology, an investor can choose, at time t1, between two capital allocations (Lagoarde-Segot, 2019).

The profitability with an un-sustainable pattern may be maximized in the short term but rate of return culminating in collapse in the longer term or a lower profit.

However, incorporating sustainability principles into financial decision-making is not so immediate. Following Schoenmaker (2017), sustainable finance (SF) can be modeled based on the level of integration of the three objectives and how some are subordinated to others.

Schoenmaker (2017) proposes framework for sustainable finance of three models: SF 1.0, profit maximization; SF 2.0, internalization of externalities to avoid risk; and SF 3.0, contributing to sustainable development while observing financial viability.

The consolidation of this new financial paradigm in its most committed version requires several key elements:

1. Redefinition of financial functions so that the allocation of resources to their most productive use flows toward economic activities and projects that are socially, environmentally, and economically sustainable.
2. Redesign of governance: This refers to the existing systems, processes, and oversight to ensure that an organization and its products meet their financial, social, and environmental objectives and to provide assurance to stakeholders that these objectives will be met.

3. Capacity for detection, evaluation, and monitoring of genuinely sustainable projects and, therefore, their proper assessment and follow-up.
4. Reformulation of risk management, integrating the so-called sustainability risks. These include risks related to the impacts of climate change, both physical and transition risks or adaptation to a new "green" economy, those associated with social considerations that economic activities must meet (e.g., reducing inequality, inclusive growth, labor relations, human rights, etc.), as well as governance factors (management structure, employee relations, executive compensation, tax and regulatory compliance, etc.). These ESG risks can translate into "traditional" financial risks such as credit or market risk, as well as nonfinancial risks, operational or reputational risks, among others.
5. A transparent and innovative financial system and financial instruments to facilitate this transition and, therefore, allocate resources based on sustainability criteria.
6. A new financial education that overcomes some of the assumptions of traditional finance.

SF thus faces significant challenges and obstacles, such as developing innovative financial instruments, increasing and managing the profitability of these tools, formulating and unifying policies and frameworks for sustainable finance, combating "greenwashing," leveraging new technologies for the consolidation of sustainable finance, and the need to modify investor behavior. Additionally, it is essential to avoid the financialization of sustainability.

Given that decisions achieving a win-win-win for the social, economic, and environmental factors are not always feasible, it is necessary to reach a trade-off that results in a benefit among these three objectives.

7.2.1 Sustainability into the Financial System

The challenges, risks, and opportunities arising from sustainability compel the financial system in general and financial institutions in particular to consider sustainability criteria in the financial decision-making process. Exposure to sustainability-related risks has been a key motivator for the financial system to adopt sustainability-related policies and develop associated products.

These sustainability risks, especially climate risk, exhibit two characteristics: they are enormously complex and multifaceted, and they can be "financial material," thus having a significant impact on all industries, companies, regions, and countries. This impact affects business models, costs and margins, revenue growth, required capital, and, of course, risk.

From the perspective of financial institutions, it is crucial to understand these risks and factor them into financial decisions because the risks associated with unsustainable development will impact not only tangible assets but also financial assets. Therefore, it is necessary to reduce the negative impact and maximize the positive social and environmental effects of financial decisions.

Financial institutions are increasingly aware that financial assets of companies engaged in unsustainable activities may become stranded assets—assets that significantly devalue due to changes in the market around them.

The reason is straightforward: companies that, under regulatory and social pressures, must pay for their negative externalities (negative impacts) will increase their external costs. This can have a domino effect, as the high external costs resulting from unsustainable activities over time can substantially reduce the ability to generate profits, as well as the value of those assets and profitability for investors.

As Schoenmaker and Schramade (2018) point out in their manual "Principles of Sustainable Finance," internalizing externalities becomes a key element of sustainable finance to reduce negative impacts and effects. A first step for companies to address or mitigate externalities is to measure and price them whenever possible because assigning financial value to social and environmental externalities facilitates the optimization process between the financial (or economic), social, and environmental dimensions.

The methodology for calculating the integrated value involves measuring, monetizing, and balancing financial and nonfinancial values, identifying the main social and environmental externalities. However, there are significant political and technological uncertainties about how these externalities will develop over time (KPMG, 2014).

Moreover, the relationship between the financial system and sustainability also arises from the potential opportunities it can create for financial institutions. There will be a demand for capital to ensure infrastructure resilience to climate change impacts and to finance the transition to a low-carbon economy, as well as opportunities to provide financial services and support in areas such as environmental technology and energy efficiency.

Furthermore, the financial sector must respond to new demands and behaviors from customers and consumers who may have concerns and demand sustainability-linked products, thereby increasing financial flows from unsustainable to sustainable activities. This ensures that the financial system can create the expected value for its customers and provide collateral benefits to the community at large, as well as avoid paying for negative externalities in the future.

All of this requires a shift in the corporate culture of financial institutions, and various initiatives in the financial sector, such as the Principles for Responsible Investment (PRI), Principles for Sustainable Insurance (PSI), Equator Principles, and UN Environment Programme Finance Initiative Principles for Responsible Banking, among others, serve as guides in the decision-making processes of these institutions.

7.2.2 Key Actors in Sustainable Finance

We will begin by providing a general overview of the main actors in the sustainable finance market and the roles they play.

Firstly, we have those seeking capital: companies, governments, international organizations, and financial entities. These organizations may issue financial assets (stocks, bonds, etc.) or borrow in the credit market to obtain financial resources. It is expected that these entities in need of funds for sustainable projects or products will be transparent about their characteristics and social and environmental impact, as well as how they align with relevant rules and regulations. In the case of companies, this connects with corporate finance and how they direct their business and behavior toward sustainable practices.

The integration of sustainable finance into corporate finance means embedding ESG considerations into every aspect of a company's financial decision-making processes. This ensures that businesses are not only financially healthy but also aligned with broader societal goals, such as reducing carbon emissions, promoting social equity, and adhering to responsible governance standards. Corporate finance decisions—ranging from capital allocation, debt issuance, and mergers to risk management and governance structures—are key levers in driving sustainable development.

Companies are increasingly recognizing that investment decisions made today must account for long-term ESG risks and opportunities to remain competitive and compliant with emerging regulations.

- Sustainable Capital Expenditures: Companies now factor sustainability into capital expenditure (CapEx) decisions, such as investing in energy-efficient technologies, renewable energy projects, or green infrastructure. For example, large manufacturers may allocate capital toward cleaner production processes or sustainable supply chain management to reduce their environmental impact and improve long-term cost efficiency.
- Sustainable Mergers and Acquisitions (M&A): Corporate finance teams must evaluate ESG risks and opportunities during mergers, acquisitions, and joint ventures. Sustainable finance ensures that companies assess the environmental and social track records of potential partners to avoid reputational risks or financial liabilities associated with unsustainable business models.

By incorporating ESG criteria into investment decisions, corporate finance teams help companies manage risks, improve operational efficiency, and position themselves to capitalize on new markets for sustainable products and services.

In a sustainable finance context, corporate finance must also account for ESG-related risks that could have long-term financial implications.

- Climate and Environmental Risks: Climate change poses significant financial risks to companies, from regulatory changes and carbon pricing to physical risks like extreme weather events. Corporate finance teams must account for these risks when evaluating long-term capital projects or financing decisions.

Companies may need to adjust their capital structures to fund climate adaptation measures, reduce carbon footprints, or comply with new environmental regulations.
- Social and Governance Risks: Poor labor practices, supply chain issues, or governance failures can lead to reputational damage, legal challenges, or financial losses. Corporate finance professionals are increasingly involved in managing these risks by ensuring that ESG factors are considered in financial planning, due diligence processes, and decision-making at the executive level.

Sustainable finance is about transforming the entire financial system to address environmental and social challenges while ensuring long-term financial stability. Corporate finance is an integral part of this transformation, as it shapes how businesses allocate capital, manage risks, and make decisions that influence sustainability outcomes. Through capital allocation, risk management, financing sustainability initiatives, and improving corporate governance, corporate finance directly supports the goals of sustainable finance.

Thus, sustainable finance cannot be fully understood or implemented without considering the role of corporate finance, which enables companies to align their financial strategies with the broader objectives of sustainability and long-term value creation. As businesses increasingly integrate ESG considerations into their financial decisions, corporate finance will remain at the forefront of driving sustainable development.

Secondly, we have investors. Investors with commitments to sustainability aim to integrate sustainability considerations into their decision-making processes to achieve satisfactory financial returns and positive impacts on sustainability. Institutional investors, specifically pension funds, insurance companies, banks, and investment funds, are currently the largest investors in sustainable financial products.

Alternative institutional investors include sovereign wealth funds, hedge funds, and private equity. As professional entities, institutional investors have the means and knowledge to engage with companies, increasing their leverage on corporates and fostering sustainable business practices. Private coalitions on sustainability can run in parallel with government initiatives and regulations on sustainability, reinforcing each other.

These institutional investors have been crucial in the development of sustainable finance, motivating many companies to strengthen their ESG management and reporting. Additionally, they have strengthened their own systems and procedures to assess sustainable projects.

Another key player is financial intermediaries that connect entities seeking funds and investors. Generally, these intermediaries are investment banks and insurers that provide savings, investment products, and other financial services to clients with an increasingly public awareness of sustainability issues. These clients demand that their money be invested or managed in a way that reflects their concerns about sustainability issues.

The mechanism facilitating the largest transactions between capital seekers and investors takes place on stock exchanges that facilitate the flow of sustainable financial capital. They also function as informers through labels or indices that help investors identify and invest in environmentally and socially responsible companies and financial products.

Financial institutions, particularly large asset managers and pension funds, have increasingly embraced shareholder activism as a tool to promote sustainable corporate governance. By using their voting power at annual general meetings (AGMs), these institutions can push companies to adopt ESG-friendly policies. This includes demanding greater transparency on sustainability issues, setting science-based climate targets, or improving diversity and inclusion within corporate leadership.

For example, investment firms such as BlackRock and Vanguard have publicly committed to voting against directors of companies that fail to address climate risks. Such engagement can take the form of direct dialogue with company executives, filing shareholder resolutions, or voting on ESG-related issues. Through these actions, financial institutions help ensure that companies prioritize long-term sustainability over short-term profit maximization.

Many financial institutions are now incorporating ESG factors into their investment decisions as part of responsible investing strategies. This integration means that companies are increasingly evaluated not just on their financial performance, but also on how well they manage ESG risks. Firms with weak ESG performance may face higher borrowing costs, limited access to capital, or even divestment by major investors.

Banks and other lending institutions can influence corporate governance through their lending practices. By offering preferential financing terms for companies that meet sustainability criteria or by requiring borrowers to adhere to specific ESG standards, financial institutions can incentivize sustainable business practices. Green loans and sustainability-linked loans, for example, are structured to reward companies for achieving certain sustainability targets, such as reducing carbon emissions or improving energy efficiency.

These financial products encourage companies to integrate ESG factors into their core operations and governance structures. Moreover, by setting strict ESG-related covenants in loan agreements, financial institutions can ensure that borrowers remain committed to sustainability objectives throughout the loan term.

Also, financial institutions, particularly those with substantial holdings in a company, can influence board composition by advocating for greater diversity and ESG expertise at the leadership level. Research shows that diverse boards are more likely to consider a wide range of stakeholder interests and address ESG concerns more effectively. By pushing for diversity in gender, ethnicity, and skills, financial institutions promote more inclusive and sustainable governance.

Stock exchanges can play a significant role in the growth of sustainable financial products. For example, they can list securities in specific sustainable segments, develop indices for sustainable financial products, and promote and share information about environmentally and socially sustainable products, making it easier for investors to find and invest in companies and products aligned with sustainability. Several stock exchanges already promote various sustainable products.

Of special importance are regulators and policymakers. In recent years, we have seen a rapid growth in the number of regulations related to sustainability adopted worldwide, as illustrated in Fig. 7.1.

From regulations, structures and rules are established to define how financial markets function and how they address concerns related to sustainability. An example is that the Principles for Responsible Investment have identified over 650 policies and regulations related to sustainable finance (SF), especially in three areas: disclosure information requirements, risk management requirements, and classes of sustainable products and assets. More information can be found at https://www.unpri.org/download?ac=12247.

Finally, external reviewers, whose role is to assess and ensure the credibility of sustainable financial products, have played a significant role in the growth of specific products such as green bonds and green loans, which we will define later.

There are different types of external reviewers (ICMA, 2018) that use various methodologies and tools. They can be grouped into:

(a) Scoring/Rating: Third parties such as rating agencies that, according to a methodology, conduct an evaluation or score that can be used for comparison with other investments.
(b) Certification: Refers to the certification of financial instruments against an appropriate standard that determines the alignment of that instrument with the established criteria to obtain certification.
(c) Verification focuses on alignment with internal or external standards established in financial instruments.
(d) Second party opinion. This option typically establishes an evaluation of a specific financial product qualified as sustainable with certain frameworks such as the green bond principles, etc.

As we can see, the sustainable financial system is a multi-actor scenario where actions from multiple stakeholders providing transparent disclosure, comparable information, adequate ratings, and standards are needed to effectively allow funding proceeds to reach sustainable ends and benefit the wider community.

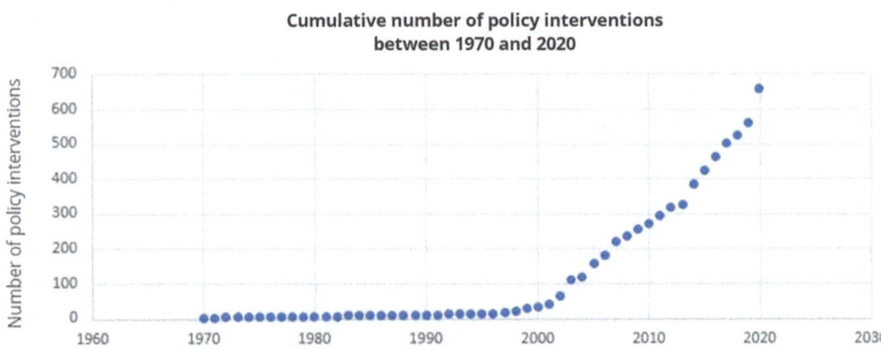

Fig. 7.1 Number of policies on sustainable finance (PRI, 2021)

7.2.3 Policy and Regulation

Many governments worldwide are developing and implementing strategies to integrate sustainability into their financial systems. In fact, China adopted the first guidelines of this kind in 2015 in the form of a guide for the issuance of green bonds and a catalog of projects backed by green bonds.

Considering that all regions see sustainable finance regulation as critical to increasing market transparency and reducing risks of greenwashing, since 2000, an increasing number of regulators and financial supervisors, central banks, and governments worldwide have been working to align financial systems with sustainable development. The key topics in sustainable finance regulation include product standards, disclosures, and labeling; management and disclosure of climate risks; management and disclosure of ESG risks; ESG in stewardship; green bond guidelines; and taxonomies (ISS ESG Report, 2022).

The implementation of regulatory initiatives varies between and within regions. Europe continues to be the leading region in the depth and breadth of regulatory initiatives on sustainable finance. Within Europe, France, the Netherlands, Germany, and Sweden have the greatest regulatory focus on sustainable finance, with a significant increase in actions seen in the UK. It is worth noting the Task Force for Climate-Related Financial Disclosures (TCFD), developed by the Financial Stability Board, to mandate climate disclosures by financial market participants. The most comprehensive regulation of this kind remains the EU's Sustainable Finance Disclosure Regulation (SFDR), which includes templates for the disclosure of pre-defined ESG factors and requires financial market participants to explain how they consider ESG risks and impacts in their investments (EU, 2022).

The Asia-Pacific (APAC) region has also significantly accelerated its regulation of the ESG investing landscape. Key countries in this region include Malaysia, Singapore, India, and Japan. Other APAC jurisdictions that have taken various regulatory initiatives include China and Hong Kong, the Philippines, Thailand, and Taiwan. Similarly, in North America, the USA has seen significantly increased interest in addressing climate risk and greenwashing through regulation. In Latin America, while several countries have green bond frameworks or ESG risk management guidelines, Brazil continues to lead the region. In Africa, South Africa continues to stand out with a broader range of actions.

All these examples manifest that policy frameworks remain inconsistent and incomplete in most countries, hindering the need to create a robust system to redirect funds to sustainable investments (UNEP FI, 2021).

7.2.4 Taxonomies

A particular focus of regulators is being placed on taxonomies to promote market transparency and integrity and channel financial flows toward sustainable activities. Sustainable finance taxonomies provide clear definitions and classifications of sustainable economic activities. They can be used to assess whether an economic activity aligns with sustainability and to what extent. They help compare investments and their sustainability, prevent greenwashing, avoid further market fragmentation created by multiple public and private sustainable financial initiatives and tools, reduce financial institutions' exposure to sustainability risks, support a more resilient financial system, and foster a more stable financial sector.

Several emerging markets and developing countries have already adopted a taxonomy, such as Bangladesh, China, Colombia, Indonesia, Mongolia, South Africa, and Sri Lanka. Additionally, two regional taxonomies have been developed: the (ASEAN, 2021) and the (EU, 2022).

Taxonomies typically include a detailed list of sectors and economic activities and a corresponding set of qualitative or quantitative criteria to determine the alignment of activities with the taxonomy's goal or goals. The basic elements of any taxonomy are objectives, sectors, activities, and criteria for taxonomy alignment.

According to the objectives, there are two types of taxonomies: green and social. Green taxonomies define activities that contribute to a country's environmental sustainability goals, usually including social safeguards to address the social impact of economic activities. Examples of green taxonomies include those of Colombia, the EU, Mongolia, and South Africa.

Social taxonomies focus on activities that positively contribute to social goals, often based on international human and social standards, principles, and objectives. In February 2022, a draft proposal for the EU social taxonomy was launched. China introduced a taxonomy of Sustainable Development Goals (SDGs) in 2020, a hybrid taxonomy that includes both environmental and social sustainability goals aligned with the SDGs.

There is also a typology called the "traffic light" system, with green, amber, and red colors, encouraging the transition from significantly harmful performance to greater alignment with taxonomy goals. The EU has outlined the planned expansion of the EU Taxonomy with a traffic light system that includes green, amber, and red activities. In contrast, the ASEAN Taxonomy contains a set of three categories (green, yellow, and red) and only specifies transition pathways for activities falling into the yellow category.

Some taxonomies may have multiple goals, which can be independent or interdependent. The interdependence of goals means that as an activity contributes to one goal, it must not cause significant harm to any other goals (DNSH, do no significant harm). When multiple sustainability goals are included in a taxonomy, the DNSH principle provides a more holistic approach to sustainability. The DNSH principle is fundamental to the taxonomies developed by the EU, Bangladesh, Colombia, and South Africa.

On the other hand, the sectors and activities covered by a taxonomy and how taxonomy alignment is defined will vary depending on the taxonomy's goal or goals and the country in which it is applied. From the taxonomy's objectives, the most relevant sectors and activities for sustainability are derived. A common mistake is to believe that economic activities included in a taxonomy are automatically considered sustainable (EU, 2022).

To determine if activities fit a taxonomy, they must be examined granularly and compared against criteria or performance thresholds. There are three common approaches to determining if an asset or activity fits a sustainable finance taxonomy: technical selection criteria, a whitelist, or principle-based criteria.

Technical selection criteria (TSC) determine the conditions under which an economic activity fits the taxonomy. The whitelist approach, as applied by China, focuses on sectors and activities contributing to the sustainability goals of the taxonomy. A key difference between principle-based taxonomies and a whitelist or TSC approach is that they do not cover specific sectors and economic activities or stipulate quantifiable thresholds for determining taxonomy alignment. Instead, they provide a set of criteria and assessment questions allowing activities to be classified based on taxonomy principles.

The principle-based approach is a less standardized approach that bases taxonomy alignment on a series of questions and assessment criteria. This approach does not include guidance on the inclusion or exclusion of specific sectors and economic activities, nor does it stipulate selection criteria to determine taxonomy alignment. The principle-based approach applied in Malaysia's Taxonomy.

China and the EU were pioneers in taxonomy development and have made significant progress since then. Both offer interesting examples of how to determine and apply taxonomy objectives, scope, and alignment. The EU Taxonomy is detailed and comprehensive, and its science-based technical selection criteria (TSC) approach has set a high sustainability standard. The taxonomy sets four conditions for an economic activity to be considered environmentally sustainable: (a) substantially contribute to one or more of the six environmental objectives; (b) not significantly harm any of the six environmental objectives (DNSH principle); (c) be carried out respecting the minimum safeguards specified by the EU; and (d) comply with the TSC established by the EU.

7.3 Sustainable Finance Products and Instruments

Investors in sustainable assets may have specific goals within the realm of sustainability. Some seek investments that empower companies leading social change, while others aim to strengthen corporate governance or enhance environmental impacts.

Some investors simply want to ensure that their investments consider the financial implications, risks, and opportunities of sustainability factors. To address these diverse needs, sustainable finance has developed instruments that can adapt to different strategies, reflecting the variety of attitudes that different types of investors have toward sustainability objectives.

Next, we will explore the main formulas for investing in sustainability.

7.3.1 An Overview of Sustainable Finance Strategy

Investment strategies express any financial investment approaches in connection with sustainability, specifically environmental, social, and governance (ESG) factors, green projects, and sustainable finance instruments.

Different strategies have been employed for sustainable investments, with the most characteristic ones including:

(a) Negative screening: This strategy incorporates moral principles into the investment decision, excluding any investment in companies involved in specific industries or activities, such as the gambling, alcohol, tobacco, or adult entertainment industries.
(b) Best in-class approach: This strategy relies on a comparison of companies, investing only in those leading in sustainability performance within their peer groups. Investors in this category include instruments that convey sustainability qualities above certain thresholds.
(c) Norms-based investment: This type of investment excludes companies and sovereign debt that do not comply with internationally accepted standards such as the UN Global Compact, UN Declaration of Human Rights, OECD, ILO, and UN norms. These norms could include, for example, the OECD Guidelines for Multinational Enterprises and UN Guiding Principles on Business.
(d) Thematic investment: This strategy involves investing in companies focused on specific aspects of sustainability, such as clean energy, green technology, or sustainable agriculture.

Additionally, there are other growing approaches that investors can adopt to integrate ESG factors into their investment practices and processes, such as ESG integration, governance impact and active ownership, and impact investment.

In the ESG integration strategy, the potential impact of ESG issues, both positive and negative, on company financials is evaluated, affecting the integration decision (EUROSIF, 2021).

In the case of governance impact and active ownership, the central feature is active engagement. Investors engage with the managers and boards of directors in all matters related to the company's sustainability policies, aiming to enhance them and thereby reduce sustainability risks and improve long-term performance.

Thus, ESG activism focuses on a broader range of stakeholders' interests, including employees, customers, and creditors (Dimson et al., 2015). It refers to the use of investors' influence over investee or potential issuing companies, policymakers, service providers, and other stakeholders to maximize overall long-term value in the context of sustainability (PRI, 2021).

Impact investments, on the other hand, aim to generate positive and measurable social and environmental impact along with a financial return (GIIN, 2020). The main characteristics of impact investment are (GIIN, 2020): intentional contribution to positive social and environmental impact, use of impact evidence and data in investment design, impact performance management, and contribution to the growth of impact investment.

The ESG strategies have experienced significant growth because, despite drastic divergences in ESG ratings regarding scope, measurement, and weighting (Berg et al., 2022), ESG investment offers more measurable and comparable metrics.

However, despite disagreements and radical differences persisting among practitioners and researchers on what qualifies as sustainable investment (Ahlström & Monciardini, 2021), the design of sustainable investments and projects to achieve their objectives must be based on five key pillars:

First, there must be a clear definition of a framework describing the sustainability characteristics and requirements of the financial product.

Second, the evaluation and selection of projects that meet the established requirements are crucial. The goal is to ensure that only assets aligning with the established framework are evaluated and selected.

The third pillar refers to traceability and tracking of funds raised for projects. These funds must be traceable and allocated in a way that demonstrates the raised money for sustainability purposes is not mixed with other sources of funding. Similarly, investment projects or assets funded by these funds should be labeled to allow traceability.

The fourth central element is reporting. Investment should inform the investor about the actual and expected impacts on the sustainability of the financed projects. Ultimately, this is about credibility, a crucial characteristic for the long-term development of sustainable financial markets.

The fifth pillar refers to verification, involving an external review conducted by independent environmental or social experts to contrast the sustainability impacts of the financed projects or activities.

Given that investments must not only provide the desired risk-adjusted financial returns but also deliver positive environmental and social impacts when investing in a sustainable financial product, it is crucial for investors to know what they are investing in and what social and environmental impacts are expected from them. This is essential for selecting and monitoring eligible investments and for reporting and verifying the impacts of these investments.

7.3.2 *Green Bonds, Green Loans, and Social Bonds and Other Ways to Do Sustainable Finance*

In November 2023, the Global Sustainable Investment Alliance (GSIA) published the sixth edition of the biennial Global Sustainable Investment Review (GSIR), revealing that a total of US$30.3 trillion is invested in sustainable assets globally (Global Sustainable Investment Alliance, 2022) (Fig. 7.2).

These financial assets, with green bonds being their primary representative, can be grouped into two main categories: assets based on use of proceeds (UoP) and those based on performance-based instruments (sustainability-linked instruments). The latter are defined as any type of fixed-income instrument whose financial and structural characteristics can vary based on whether the issuer achieves social, environmental, or governance objectives.

In general, green, social, and sustainability bonds fall into the first category. They are defined as any type of fixed-income instrument in which the proceeds from their issuance are exclusively allocated to eligible environmental and social projects or a combination of both. While green bonds'proceeds are used to finance environmental projects, social bonds' proceeds fund social projects such as economic inclusion, gender equality, and education, covering aspects like health, employment, gender equality, affordable housing, etc.

Sustainability bonds, on the other hand, aim to combine both green and social benefits into a single instrument. Additionally, the category may include transition bonds with UoP earmarked for activities that are not low- or zero-emission (i.e., not green) but play a role in decarbonizing an activity or supporting an issuer in transitioning to Paris Agreement alignment in the short or long term. The transition label allows for the inclusion of a more diverse set of sectors and activities, including

	Green	Social	Sustainability*	Transition	SLB
Total size of market (cumulative)	USD2.2tn	USD653.6bn	USD682.0bn	USD12.5bn	USD204.2bn
Number of issuers	2,457	772	507	39	336
Number of countries	85	49	57	12	50
Number of currencies	49	42	41	7	21

Fig. 7.2 Size of bond market (Sustainable Debt Global State of the Market 2022 Climate Bonds Initiative)

labels like blue transition and green transition. Currently, transition bonds primarily originate from industries that are challenging to eliminate, such as extractive industries like mining, materials like steel and cement, and industrial sectors like aviation and maritime transport.

In 2007, the European Investment Bank and the World Bank issued the Climate Awareness Bond, the world's first climate protection bond. In 2008, the World Bank issued the first green bond specifically labeled for institutional investors, establishing a methodology that laid the foundation for green bond principles. Since then, the market has experienced substantial growth.

Initially, major issuers were supranational organizations, such as multilateral development banks. Over time, a variety of entities have entered the green bond market, including financial corporations, nonfinancial companies, and development banks, which currently constitute the three main issuance categories.

Regarding sectors, Energy, Buildings, and Transport have consistently been the three largest use of proceeds (UoP) categories, collectively contributing 77% of the total green debt volume, as illustrated in Fig. 7.3.

In 2014, a consortium of investment banks established voluntary best practice guidelines known as the "green bond principles" (GBP). The founding members included Bank of America Merrill Lynch, Citi, Crédit Agricole Corporate and

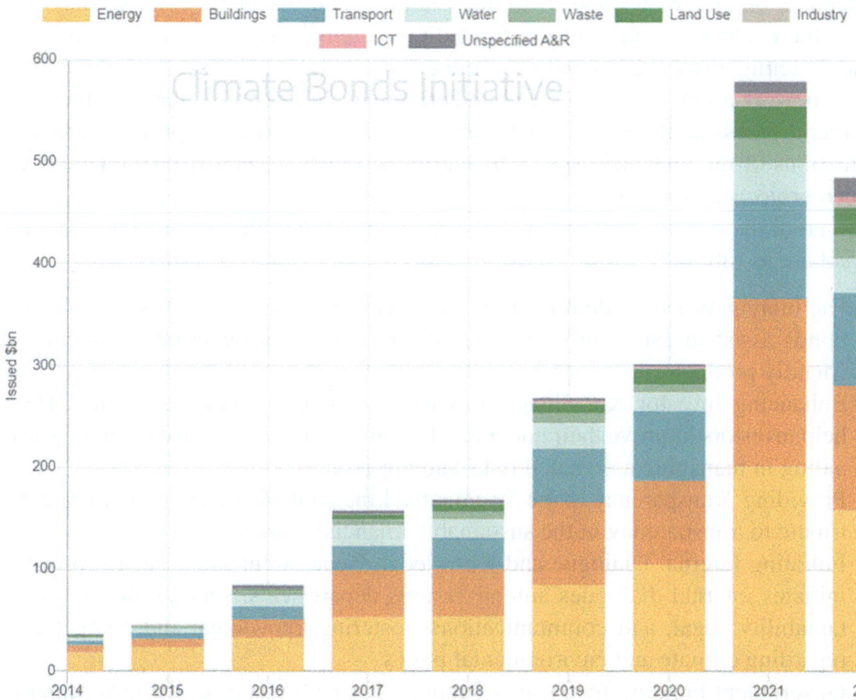

Fig. 7.3 Green debt volume per sector (Sustainable Debt Global State of the Market 2022 Climate Bonds Initiative)

Investment Bank, JPMorgan Chase, BNP Paribas, Daiwa, Deutsche Bank, Goldman Sachs, HSBC, Mizuho Securities, Morgan Stanley, Rabobank, and SEB. Since then, the oversight and development of the guidelines have been transferred to an independent secretariat within the International Capital Market Association (ICMA).

The principles are a set of voluntary guidelines that provide issuers with guidance on key components involved in launching a credible green bond. The GBP emphasize transparency, accuracy, and integrity of the information that issuers will disclose and communicate to stakeholders. The GBP consists of four basic components:

(a) Use of proceeds
(b) Project evaluation and selection process
(c) Management of proceeds
(d) Reporting

The green bond principles encourage transparency in the use of proceeds and recommend that an issuance be assessed by an external reviewer. This could be, for example, an independent second opinion provider, an external auditor, a green bond certifier, or a credit rating agency.

This independent external review provides investors with confidence in the information provided by the issuer regarding the environmental impact and procedural credentials of the green bond. It is important to note that these reviews may not assess the ecological nature of the funded projects but may focus on whether the issuer is acting in accordance with the framework defined for its issuance or whether the framework aligns with the green bond principles. Furthermore, these reviews are not credit assessments of the issuer, as credit rating agencies separately provide evaluations of the financial aspects based on the issuer's credit risk profile and all funded projects and assets.

Some of the benefits that green bonds can contribute to the sustainable financial market are as follows:

(a) Facilitating Investor Identification and Investment in Green Products: Green bonds assist investors in easily identifying and investing in environmentally friendly products.
(b) Enhancing Investor Knowledge on Climate and Environmental Issues: They help investors improve their understanding of climate and environmental issues, aiding in managing associated risks and opportunities.
(c) Providing Transparency to the Sustainable Financial Market: Green bonds contribute to transparency in the sustainable financial market.
(d) Initiating Internal Dialogue and Knowledge Generation: The issuance process initiates internal dialogues among various departments, such as finance, sustainability, legal, and communications, fostering knowledge and competence regarding climate and environmental issues.
(e) Government Funding for Project Support: Through green bonds, governments can raise funds to finance select projects and support the achievement of sustainability goals.

However, some disadvantages can be noted, such as limiting the availability of capital for projects or activities not classified as green. There is also a perception that green investment is a form of investment like conventional investments, potentially favoring the financialization of sustainability. Lack of differentiation between projects meeting the criteria, along with market confusion due to different standards and frameworks, are other drawbacks to consider.

On the other hand, green loans can be defined as loans (no asset issuance) where the borrowed money is used for environmental purposes. Typically provided by financial entities, particularly banks, these loans can be funded like any other loan or through the proceeds of a previously issued green bond, as not all borrowers can turn to the capital market for bond issuance.

This instrument is suitable for organizations or companies with financing needs too small to access the capital market, playing a crucial role in enabling and financing the transition to a more sustainable economy. Green loans are often available for small and medium enterprises (SMEs) and individuals, particularly when their capital markets are too immature.

This accessibility allows loan providers to interact with these businesses to raise awareness of sustainability issues. The provision of green loans can also offer significant business opportunities for banks, leading them to develop expertise in sustainability.

Although there is no authority deciding on the definition of green loans, there are voluntary principles developed by the Loan Market Association (Loan Market Association, 2021). As with green bonds, the choice of green loans should start by developing a green loan framework outlining the intended use of proceeds, project selection process, revenue management, and reporting on environmental impacts. This ensures transparency, integrity, and internal governance for effective product monitoring.

7.3.3 Assessing the Impact of Investments on Sustainability Goals

Performance-Based Instruments

Sustainability-Linked Bonds (SLBs) or Sustainability-Linked Loans (SLL): These are forward-looking, performance-based debt instruments wherein progress toward entity-level key performance indicators (KPIs), such as GHG emission reductions, is measured by sustainability performance targets (SPTs) within a specified timeframe. When used credibly and ambitiously, SLBs can serve as powerful transition finance instruments for issuers, demonstrating to investors their commitment to transition plans (Climate Bonds Initiative, 2022).

It is noteworthy that currently, the market for sustainability-linked loans is significantly larger than that of sustainability-linked bonds. In fact, between 2017 and 2021, the volume of sustainability-linked loans issued was approximately ten times the volume of sustainability-linked bonds issued.

Let's begin by explaining the difference between the use of proceeds model and the performance-based model. As we have seen, the use of proceeds model entails the issuer or borrower allocating funds for green, social, or sustainable projects or assets. However, the performance-based model involves the issuer or borrower committing to future improvements in sustainability-related outcomes in their operations. Sustainability-linked bonds or loans encompass any type of bond instrument whose financial and/or structural features, such as coupon maturity or redemption amount, may vary depending on whether the issuer achieves predefined sustainability, environmental, social, or governance objectives within a predetermined timeline.

These are not the only differences; for instance, the use of proceeds models focuses on specific investments or assets, whereas performance-based models focus on broader, more general issuer or borrower activities toward sustainability.

Another distinction is that performance-based models involve changes in financial characteristics, such as interest rates, based on performance. These instruments must be linked to specific key performance indicators (KPIs) and sustainability performance targets (SPTs), so financial conditions will change depending on whether sustainability results are achieved or not.

Financial institutions have also developed a set of voluntary principles. Sustainability-linked bond principles are a set of voluntary guidelines providing issuers with guidance on key components involved in the issuance of a sustainability-linked bond, based on five elements: KPI selection, SPT selection, financial instrument characteristics, reporting, and verification.

This entails clearly defining KPIs and SPTs in the bond documentation, along with a clear specification of how bond financial characteristics will alter based on performance. Any exceptions, such as situations beyond the issuance's control, must be identified in the bond documentation.

Verification reports on SPT performance should seek independent external assessment of their performance against each SPT at least once a year, and this information should be publicly available.

Linked loans function very similarly to sustainability-linked bonds. In general, the benefits of these instruments are highly relevant, as they demonstrate the issuer's real sustainability strategy and commitment to their overall activities and business model, rather than a specific project. Additionally, these instruments are more flexible than the use-of-proceeds model, given their general purpose.

However, a holistic understanding of sustainable finance goes beyond a narrow focus on specific financial products, such as green bonds or ESG investments. Instead, it encompasses the transformation of the entire financial system, with the aim of aligning capital flows with the goals of sustainable development, economic stability, and social well-being. To understand its full impact, it is crucial to explore how sustainable finance influences different sectors of the economy, affects social structures, and transforms business and government strategies.

Sustainable finance directly impacts the broader economy through its influence on investment patterns, resource allocation, and long-term growth strategies. As more investors prioritize ESG factors, companies that perform well in sustainability metrics gain preferential access to capital, leading to several economic implications.

First, sustainable finance directly impacts on resource efficiency and innovation. Sustainable finance incentivizes companies to adopt more resource-efficient practices and invest in innovation. For instance, corporations that seek green financing may be required to demonstrate progress in reducing carbon emissions, waste, or water usage. This shift encourages innovation in cleaner technologies, renewable energy, and sustainable supply chains, ultimately driving productivity and enhancing the long-term competitiveness of economies.

Second, risk mitigation and financial stability: Climate change, social unrest, and poor governance are increasingly recognized as systemic risks to financial markets. By integrating ESG factors, sustainable finance seeks to mitigate these risks, thus contributing to the stability of financial systems. For example, the risk of stranded assets—fossil fuel reserves that may become obsolete due to climate regulation—has become a focal point for investors concerned about climate transition risks. Sustainable finance, therefore, promotes resilience against economic shocks that could arise from environmental or social crises.

Third, SF affects capital reallocation and sectoral shifts: As capital flows toward sustainable investments, sectors that align with sustainability goals, such as renewable energy, electric vehicles, and circular economy initiatives, are likely to thrive. Conversely, carbon-intensive industries, such as oil and gas or traditional mining, may face declining investment and higher costs of capital. This reallocation of resources has the potential to reshape entire industries, stimulate job creation in emerging sectors, and foster a more sustainable economic structure.

To finalize, economic inclusivity: SF also addresses issues of social equity by encouraging investments in social infrastructure, such as affordable housing, healthcare, and education. Social bonds, for example, are specifically designed to finance projects that generate positive social outcomes, such as reducing income inequality or improving access to essential services. By promoting investments that benefit marginalized communities, sustainable finance supports a more inclusive and equitable economic model.

In addition to the economic implications, sustainable finance can also have important social effects. Sustainable finance's impact on society extends beyond economics to address crucial social challenges, including inequality, human rights, and corporate accountability. As financial markets increasingly integrate ESG considerations, companies are pressured to adopt more responsible practices, with wide-reaching social implications.

The emphasis on ESG has intensified corporate efforts to demonstrate their commitment to social responsibility. Companies are now evaluated on their labor practices, community engagement, diversity and inclusion policies, and the ethical management of their supply chains. SF encourages businesses to adopt transparent and fair labor practices, promote gender equality, and engage in community development projects. These efforts not only improve corporate reputations but also foster healthier relationships between businesses and the societies in which they operate.

Socially responsible investing (SRI) and impact investing are designed to tackle inequality by channeling capital into projects that directly address social disparities.

Initiatives such as microfinance, affordable housing, and sustainable agriculture can empower low-income communities, support small-scale entrepreneurs, and reduce the wealth gap. In this way, sustainable finance promotes social mobility and empowers marginalized groups by creating opportunities for financial inclusion and economic empowerment.

On the other hand, local and regional development can benefit significantly from sustainable finance. Green bonds, for instance, have been used to fund projects that improve public infrastructure, enhance public transportation, and reduce urban pollution. Sustainable finance encourages partnerships between the private sector, governments, and communities to achieve common goals that improve the quality of life and promote resilience against environmental and economic challenges.

It is also important to note that consumers and investors alike are increasingly aware of the ethical dimensions of financial decisions. SF amplifies consumer and investor demands for businesses to act responsibly. By directing capital away from companies engaged in harmful practices, such as child labor or environmental degradation, sustainable finance helps to raise the bar for corporate conduct across industries. As consumer preferences for ethical products grow, companies face stronger incentives to align their business models with sustainable principles.

While challenges remain—such as the need for more stringent regulations, overcoming greenwashing, and expanding access to sustainable finance—its potential to reshape economies and societies is immense. A holistic view of sustainable finance recognizes that its ultimate goal is not just financial returns, but the creation of a more just, equitable, and sustainable world for future generations.

7.4 Evaluating the Effectiveness of Impact Measurement Methodologies

As previously discussed, one of the elements that can solidify sustainable finance is its ability to assess and monitor the outcomes of sustainable projects/activities/assets. However, in a context of heterogeneous regulations and an immense need for funds to achieve sustainability goals, a fundamental barrier emerges. There is no common framework for assessing the impact of sustainable investments, undermining investor confidence in sustainable financial markets and products (Berg et al., 2022).

Several factors contribute to this situation. Firstly, the plethora of methods and providers for evaluation, along with different tools or methodologies, fails to fully reflect the direct contribution to sustainability. Secondly, the results of methodologies often differ significantly and lack consistency. These variations arise from different methodologies originating from various providers, including independent institutions, academics, and third-party data providers, including labeling agencies. These methodologies can be broadly categorized into two main blocks (Popescu, et al., 2021):

A. Methods focused on climate impact

B. Methods focused on measuring social and environmental impact

Within each approach, various levels exist, employing different methodologies and tools:

- Focus on climate change impact
  - *Level 1*
    - Carbon footprints and exposure metrics
      - *Level 2*
        - Carbon footprint and intensity metrics using reported and estimated data
        - Input-output-based carbon footprints
    - Alignment with low-carbon pathways
- Focus on social and environmental impacts
  - *Level 1*
    - ESG ratings
      - *Level 2*
        - Proprietary ESG ratings and scores
        - Free-to-search ratings
    - Sustainability labels
    - Sustainability-based impact assessments
      - *Level 2*
        - SDG-based assessments
        - LCA-based assessments
        - Other third-party environmental and social indicators

Determining the best method for assessing impact in sustainable investments is a complex task, as highlighted by the OECD in 2021. Progress in this area involves establishing the appropriateness of various methodologies and tools, necessitating the definition of relevant criteria for their evaluation. However, there is a lack of clear and standard criteria for making quick comparisons. Recent research by Popescu (2021) identifies seven criteria representing a potential consensus from stakeholders in the sustainable investment market:

- Double Materiality: Considering sustainability impacts from both organizational perspectives—inside out (enterprise's material impacts on sustainability) and outside in (material sustainability impacts on the enterprise)
- Reliability: Refers to the qualities of being easy to understand, accessible, and adaptable to different financial products

- Life Cycle Consideration: Stems from sustainability science, emphasizing the consideration of impact over the complete life cycle of all activities/products underlying held companies
- Comprehensiveness of Impact Categories: The extent to which methods capture both positive and negative impacts over the long term and across social, environmental, and economic areas
- Compatibility with Science-Based Targets (SBTs) for Sustainable Development: The ability of an assessment tool to frame impact in terms of investment contribution to defined sustainability, enabling meaningful comparisons
- Prospectiveness: Dynamic representation of how investments contribute to achieving longer-term advancements beyond a specific sustainability goal
- Investor's Additionality: The contribution of the investor that wouldn't have been achieved without their participation

In summary, the significance of these criteria lies in ensuring that impact assessment methods consider various perspectives, are reliable, encompass life cycle considerations, capture a comprehensive range of impact categories, align with science-based targets, exhibit dynamism, and highlight the additional value brought by investors.

Despite these criteria, recent research indicates that no existing method fully meets all of them, presenting mixed results across different criteria. This underscores the ongoing challenge of developing a universally accepted and comprehensive framework for evaluating the impact of sustainable investments.

However, despite the rapid growth of ESG investing—now accounting for trillions of dollars in global assets—there is increasing scrutiny over whether ESG strategies are delivering on their promises. In many cases, ESG approaches have failed to produce the intended real-world impact, raising questions about their effectiveness. In this text, we will critically examine why some ESG strategies have fallen short, exploring key areas of failure, underlying causes, and lessons that can be learned to improve future approaches.

ESG criteria are often applied inconsistently across industries, markets, and regions, leading to confusion about what constitutes a "sustainable" or "responsible" investment. This lack of clarity can result in ESG strategies that do not deliver meaningful impacts because they fail to adequately target or measure the desired outcomes.

- Greenwashing: Without clear definitions, some companies or investment products may engage in greenwashing—misleadingly presenting themselves as more sustainable than they are. For example, a company might emphasize minor environmental initiatives while continuing harmful practices elsewhere in its operations. This disconnect undermines the credibility of ESG approaches and results in investments that fail to produce real-world environmental or social benefits.
- Inconsistent ESG Ratings: Another issue is the inconsistency of ESG ratings from various providers. Different agencies use different methodologies, leading to conflicting scores for the same company.

ESG strategies have largely focused on individual companies rather than addressing the broader systemic issues that contribute to unsustainable practices. While ESG investment might encourage a specific firm to improve its performance on certain metrics, it often fails to address the structural factors driving environmental degradation or social inequality at a global level. Many ESG funds claim to promote climate-friendly practices, yet global carbon emissions have continued to rise. One reason for this is that many ESG funds hold shares in companies within carbon-intensive industries, such as energy or manufacturing, even if those companies score well on ESG metrics like governance or social responsibility. This limited focus on reducing absolute emissions across sectors means that ESG strategies often fail to drive the large-scale changes needed to address climate change.

Another fundamental flaw in many ESG strategies is their emphasis on short-term financial performance rather than long-term sustainability outcomes. Many ESG funds and companies prioritize maintaining competitive returns for investors over driving substantive ESG improvements. This short-term focus can dilute the potential for real-world impact, as sustainability issues—particularly climate change and social inequality—require long-term planning and investment.

By addressing the shortcomings in ESG strategies—such as inconsistent standards, short-termism, and ineffective engagement—financial institutions and investors can help ensure that ESG investing lives up to its promise of driving positive environmental, social, and governance outcomes. Moving forward, ESG must become more focused, transparent, and impactful if it is to truly contribute to a sustainable global economy.

7.5 Summary

This chapter has discussed the concept of sustainable finance, exploring its different dimensions to analyze its connection with sustainable development and the potential challenges that may arise. Additionally, since sustainable finance represents a new financial paradigm that goes beyond the pursuit of financial profitability, we have briefly reviewed its foundations and its integration into the financial system.

We highlighted the roles of various stakeholders, the policies and regulations supporting them, and the prominent financial instruments available for sustainable finance.

Finally, we have pointed out one of the main challenges facing sustainable finance: improving the transparency and integrity of the sustainable financial market with consistent evaluation methods.

7.6 Discussion Questions

1. Discuss the potential contribution that sustainable finance can make to achieving the SDGs.
2. Why do investors consider sustainability factors while allocating capital?
3. Explore in greater depth the role of financial institutions as key players in sustainability. Can they hinder sustainability?
4. Discuss whether organizations committed to responsible investments have lower rates of return.
5. Evaluate whether a specific bond aligns with the requirements of the green bond principles, the social bond principles, or the sustainability-linked bond principles.

References

Ahlström, H., & Monciardini, D. (2021). The regulatory dynamics of sustainable finance: Paradoxical success and limitations of EU reforms. *Journal of Business Ethics*, 1–20.

Aracil, E., & Sancak, I. (Eds.). (2023). *Essential concepts of sustainable finance: An AZ guide*. Taylor & Francis.

ASEAN (2021) Taxonomy for Sustainable Finance, Version 1. Retrieved from https://asean.org/wp-content/uploads/2021/11/ASEAN-Taxonomy.pdf#page=41

Berg, F., Koelbel, J. F., & Rigobon, R. (2022). Aggregate confusion: The divergence of ESG ratings. *Review of Finance, 26*(6), 1315–1344.

Cunha, F. A. F. D. S., Meira, E., & Orsato, R. J. (2021). Sustainable finance and investment: Review and Research agenda. *Business Strategy and the Environment*. https://doi.org/10.1002/bse.2842

Dimson, E., Karakaş, O., & Li, X. (2015). Active ownership. *The Review of Financial Studies, 28*(12), 3225–3268.

EU Platform for Sustainable Finance. (2022). *The extended environmental taxonomy: Final report on taxonomy extension options supporting a sustainable transition*. Retrieved from https://commission.europa.eu/system/files/2022-03/220329-sustainable-finance-platform-finance-report-environmental-transition-taxonomy_en

EUROSIF. (2021). *Responsible investment strategies*. Retrieved from https://www.eurosif.org/responsible-investment strategies/

Global Impact Investing Network (GIIN). (2020). *Core characteristics of impact investing*. https://thegiin.org/assets/Core%20Characteristics_webfile.pdf. Global Sustainable Investment Alliance.

Fullwiler, S. (2016). Sustainable finance: Building a more general theory of finance. In *Routledge handbook of social and sustainable finance* (pp. 17–34). Routledge.

Global Sustainable Investment Alliance (2022). *Global Sustainable Investment Review*. https://www.gsi-alliance.org/wp-content/uploads/2023/12/GSIA-Report-2022.pdf

Gupta, J., & Vegelin, C. (2016). Sustainable development goals and inclusive development. *International Environmental Agreements: Politics, Law and Economics, 16*, 433–448.

Hahn, T., Figge, F., Pinkse, J., & Preuss, L. (2018). A paradox perspective on corporate sustainability: Descriptive, instrumental, and normative aspects. *Journal of Business Ethics, 148*, 235–248.

ISS ESG Report Climate Bonds Initiative. (2022). *Sustainable Debt Global State of the Market* (2022). Retrieved from https://www.lma.eu.com/application/files/9115/4452/5458/741_LM_Green_Loan_Principles_Booklet_V8.pdf

ICMA. (2018). *Green bond principles*. Retrieved from https://www.icmagroup.org/sustainable-finance/the-principles-guidelines-and-handbooks/green-bond-principles-gbp. Depth & Breadth of Sustainable Finance Regulatory Initiatives: Global Developments in 2022.

KPMG. (2014). *A new vision of value: Connecting corporate and societal value creation Amsterdam*. https://assets.kpmg.com/content/dam/kpmg/pdf/2014/10/a-new-vision-of-value-v1.pdf%2D%2D

Kumar, S., Sharma, D., Rao, S., Lim, W. M., & Mangla, S. K. (2022). Past, present, and future of sustainable finance: Insights from big data analytics through machine learning of scholarly research. *Annals of Operations Research*, 1–44.

Lagoarde-Segot, T. (2019). Sustainable finance. A critical realist perspective. *Research in International Business and Finance, 47*, 1–9.

Loan Market Association (2021). *Green loan principles*. Retrieved from https://www.lma.eu.com/application/files/9716/1304/3740/Green_Loan_Principles_Feb2021_V04.pdf

Nilsson, M., Chisholm, E., Griggs, D., Howden-Chapman, P., McCollum, D., Messerli, P., & Stafford-Smith, M. (2018). Mapping interactions between the sustainable development goals: Lessons learned and ways forward. *Sustainability Science, 13*, 1489–1503.

Popescu, I. S., Hitaj, C., & Benetto, E. (2021). Measuring the sustainability of investment funds: A critical review of methods and frameworks in sustainable finance. *Journal of Cleaner Production, 314*, 128016.

PRI. (2021). *Why the PRI is working towards a sustainable financial system*. https://www.unpri.org/sustainable-financialsystem/why-the-pri-is-working-towards-a-sustainable-financial-system/198.article

Schoenmaker, D. (2017). *Investing for the common good: A sustainable finance framework* (p. 80). Bruegel.

Schoenmaker, D., & Schramade, W. (2018). *Principles of sustainable finance*. Oxford University Press.

Schoenmaker, D., & Schramade, W. (2019). Investing for long-term value creation. *Journal of Sustainable Finance & Investment, 9*(4), 356–377.

UNEP FI. (2021). *A legal framework for impact*. Retrieved from https://www.unepfi.org/legal-framework-for-impact

United Nations. (2015). *General Assembly of the United Nations in September 2015*. https://sustainabledevelopment.un.org/sdgs

Yan, S., Ferraro, F., & Almandoz, J. (2019). The rise of socially responsible investment funds: The paradoxical role of the financial logic. *Administrative Science Quarterly, 64*(2), 466–450.

Chapter 8
ESG Reporting and Communication

Michaela Bednárová

Abstract In this chapter, we explore ESG reporting and communication. Students will be first introduced to global ESG reporting frameworks and standards, focusing mostly on Global Reporting Initiative (GRI) Standards, which are the most widely implemented globally for sustainability reporting. The concept of integrated reporting (IR) also deserves to be addressed in-depth as it is used in International Financial Reporting Standards (IFRS). Other relevant standards that will be outlined in this chapter include SASB, which are of major importance around the world, and SGDs—since they are part of Agenda 2030.

Alignment initiatives, such as ISSB and CSRD regulation, which introduce the European Sustainability Reporting Standards (ESRS) will also be referred to in this section, and we will briefly mention other relevant reporting initiatives such as Task Force on Climate-related Financial Disclosures (TCFD), Principles for Responsible Investments (PRI), and the Carbon Disclosure Project (CDP).

We will also review and discuss current communication strategies for conveying ESG performance and impact and address ethical reporting considerations including concerns such as greenwashing and bluewashing.

Additionally, corporate digital responsibility (CDR) will be introduced as a new dimension of corporate social responsibility and sustainability. In this way, a new approach to ESG reporting has emerged reflecting digitalisation in terms of the use of new technologies, artificial intelligence, etc.: ESG thus becomes ESGD, where D stands for digitalisation.

M. Bednárová (✉)
Department of Financial Economics and Accounting, University Pablo de Olavide, Seville, Spain
e-mail: mbed@upo.es

M. Bednárová, K. Soratana (eds.), *Environmental, Social, and Governance (ESG) Investment and Reporting*, https://doi.org/10.1007/978-3-031-84235-1_8

8.1 Introduction to Global ESG Reporting Frameworks and Standards

ESG reporting consists of going beyond a purely economic or financial corporate perspective and disclosing information on a company's environmental, social, as well as governance performance and impact. As economic aspects are financial in nature, the ESG approach is referred to as non-financial disclosure. Nowadays, companies all around the world use ESG reports to share information on their performance and impact on a wide range of sustainability topics including, but not limited to, CO_2 emissions, water consumption, environmental footprint, materials and resource use, waste management, supply chain issues, social and governance aspects, etc. (Roca & Searcy, 2012). Thus, different stakeholders such as investors, governments, NGOs, and others can use ESG disclosure to assess a company's impact on the environment as well as society and make informed decisions.

The ESG approach is embedded into corporate social responsibility, a concept introduced by Bowen (1953), who is regarded as one of the fathers of CSR. He defined CSR as follows: "It refers to the obligation of businessmen to pursue those policies, to make those decisions, or to follow those lines of action which are desirable in terms of objectives and values of our society". Similarly, Carroll's (1991) CSR pyramid outlines the four main types of responsibility of a business: economic (being profitable); legal (obeying laws and regulations); ethical (doing what is fair and avoiding harm); and philanthropic (being a good corporate citizen). Thus, CSR goes beyond a company's legal obligations and refers to its commitment to operating not only in an economically sustainable manner but also in a socially and environmentally responsible way. This brings us closer to the broader concept of sustainability, which encompasses both the ESG approach and CSR. Sustainability has been defined as an ability to meet the needs of the present without compromising the ability of future generations to meet their own needs. Hence, sustainability and CSR go hand in hand as a sustainable future is not possible without company awareness and accountability regarding the obligation of being a responsible corporate citizen (UN, 2015). ESG and sustainability are not synonyms. Sustainability is a broader concept and emphasises business sustainability, which is not possible without considering economic aspects. We can thus understand sustainability as an EESG approach, which stands for economic, environmental, social, and governance.

Nowadays, ESG disclosure is integrated into corporate reporting, and corporate reports dealing with these matters come under different titles: ESG report, non-financial report, green report, etc. When combined with economic aspects (EESG), we call them sustainability reports or integrated reports.

ESG began to attract serious corporate interest in the 1990s, and since then, CSR disclosure has been mostly voluntary. Only recently has CSR been established as a requirement by some governments and stock exchanges. Academics have thus attempted to explain the reasons driving companies to engage in voluntary ESG reporting and have proposed theories, among which *legitimacy theory* is one of the most advanced and accepted. According to this theory, companies voluntarily disclose information about environmental and social issues to legitimise their activities and to prove that they are in line with societal expectations (Suchman, 1995). Similarly, *stakeholder theory* (Freeman, 1984) suggests that an organisation's effectiveness is measured by its ability to satisfy not only shareholder interests but also different stakeholders, pointing to the relevance of non-financial transparency.

We have recently witnessed big shifts in non-financial disclosure as regulators and standard-setters around the world have taken significant steps forward in this domain. In March 2021, Sustainable Finance Disclosure Regulations (SFDR) were implemented in the EU. In November 2021, the Hong Kong Stock Exchange published mandatory climate disclosure guidance. At the same time, the IFRS foundation announced the formation of its global reporting standardisation initiative. A month later, in December 2021, the Singapore Stock Exchange issued mandatory disclosure recommendations on climate and board diversity. In March 2022, the US Security Exchange Commission (SEC) set out a climate disclosure proposal. In June 2022, the Guidance for Enterprise ESG Disclosure took effect in China. At the same time, the Johannesburg Stock Exchange published guidance documents on voluntary sustainability and climate change disclosure (KPMG, 2022).

The most recent sustainability reporting KPMG survey shows that 96% of G250[1] companies report on sustainability or ESG matters (Fig. 8.1); 64% acknowledge climate change as a risk to their business; and 49% acknowledge social elements as a risk to their business. The most commonly used anchors in recent ESG and sustainability reporting are the *Global Reporting Initiative (GRI)*; the *Task Force on Climate-related Financial Disclosures (TCFD)*; and the *United Nations Sustainable Development Goals (SDGs)* (KPMG, 2022).

8.1.1 Global Reporting Initiative (GRI)

GRI is an international non-profit organisation with a network-based structure including thousands of professionals and organisations across many sectors and countries. The GRI's Secretariat is located in Amsterdam with focal points in Australia, Brazil, China, India, South Africa, and the USA.

The Global Reporting Initiative was founded in 1997 by the Coalition for Environmentally Responsible Economies (CERES) and the Tellus Institute with the support of the United Nations Environment Program (UNEP). It is directed towards promoting sustainable development, and the initiative has pioneered and developed a comprehensive sustainability reporting framework. The latter constitutes today the world's most commonly used reporting standards. GRI standards offer a flexible framework for preparing standalone sustainability or non-financial reports, as well as integrated ESG reports. GRI enables all organisations to measure and report their

[1] The G250 refers to the world's largest 250 companies by revenue based on the 2021 Fortune 500 ranking.

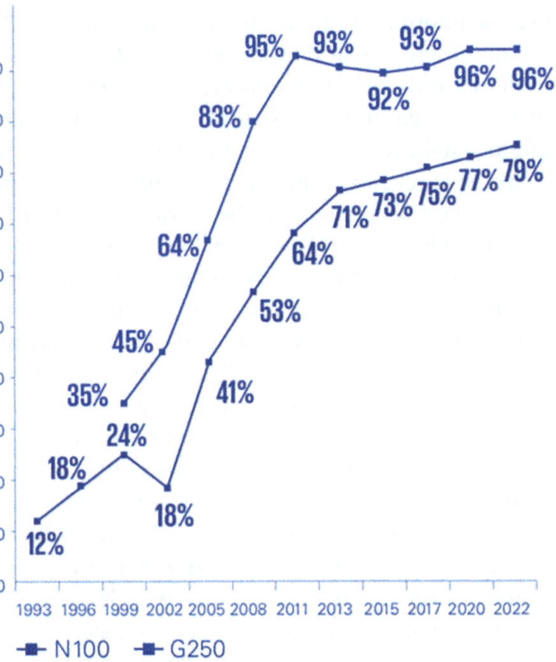

Fig. 8.1 Global sustainability reporting rates (KPMG, 2022)

economic, environmental, and social performance, i.e. their impact in three key areas of sustainability. The sector guidance and sustainability reporting guidelines were also launched to make it easier for companies to adopt them.

A wide range of stakeholders are interested in the transparency of a company's sustainability issues: businesses, investors, accountancy, labour, non-governmental organisations, etc. That is why GRI has relied on collaborations and consensus-seeking consultations of a large network of experts from different fields and stakeholder groups. This multi-stakeholder approach has given widespread credibility to the GRI Sustainability Framework.

GRI has formed a global strategic partnership with well-known and internationally recognised institutions such as the Organisation for Economic Co-operation and Development (OECD); the United Nations Global Compact; the United Nations Environment Programme (UNEP); the International Organization for Standardization (ISO); the Earth Charter Initiative; the International Financial Corporation (IFC); and the United Nations Conference on Trade and Development (UNCTAD).

Standards Development Milestones

The first version of the GRI sustainability guidelines was launched in 2000. It rapidly became the leading global sustainability reporting system. By August 2002, the so-called G2, the second edition of the guidelines, was released in Johannesburg during the World Summit on Sustainable Development. The GRI governance structure was completed by the end of 2005, and the third generation of guidelines, G3, was released in October 2006. The next generation, the G3.1 guidelines, was launched in 2011 and included additional guidance for reporting on human rights, local community impacts, and gender. In addition, it introduced the technical protocol—that is, guidance on how to define the content of a sustainability report. In May 2014, the GRI launched its fourth generation of guidelines, the G4 Framework, including new and revised disclosures regarding the supply chain, governance and remuneration, ethics and integrity, anti-corruption and public policy, emissions, and energy. In October 2016, the *GRI Standards*, which has been updated and developed via a robust multi-stakeholder process, replaced the G4 version. The latest version of GRI Standards was available as of 2021. The GRI Standards can be used by any organisation (public or private) from any sector including small companies.

Structure of the GRI Standards

As shown in Fig. 8.2, the GRI Standards represent a modular system of interconnected standards comprising three series of Standards: the *GRI Universal Standards* (applicable to all organisations); the *GRI Sector Standards* (applicable to specific sectors); and the *GRI Topic Standards* (including disclosures relating to a particular topic). The standards are designed to guide organisations in determining the topics that constitute material (i.e. which topics are relevant to report on) (GRI, 2021).

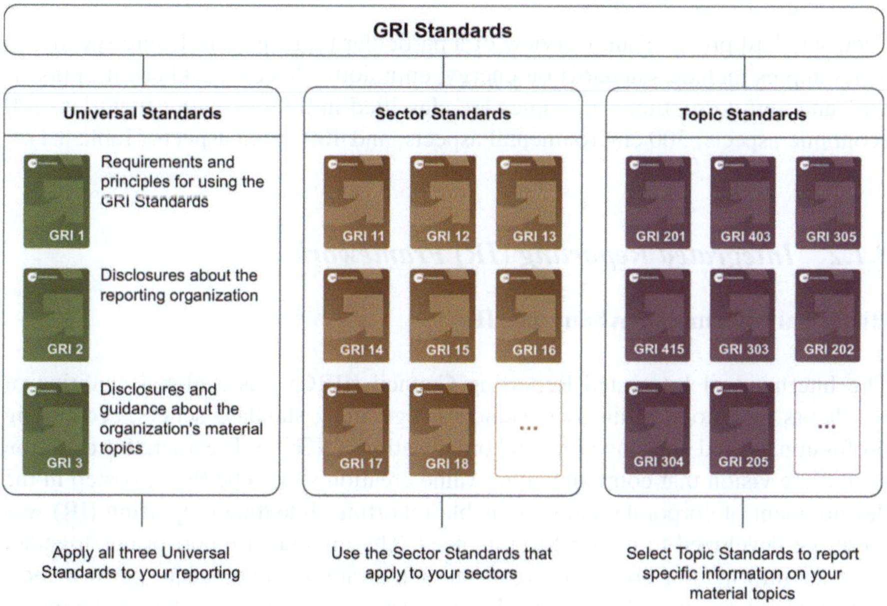

Fig. 8.2 GRI Standards structure (GRI, 2021)

GRI Universal Standards

Universal Standards are further broken down into *GRI 1: Foundation 2021*; *GRI 2: General Disclosures 2021*; and *GRI 3: Material Topics 2021*. GRI 1 presents reporting principles such as accuracy, balance, and verifiability. It clarifies the main concepts, explains how to use the Standards, and lists the minimum requirements that need to be fulfilled. GRI 2 guides organisations on how to report about aspects such as the organisation's structure, activities, governance, strategy, employees, policies, and practices to ensure a better understanding of an organisation's impact. GRI 3 explains how organisations can identify topic material and outlines how the Sector Standards are used in this process.

GRI Sector Standards

The Sector Standards will cover 40 different sectors and are designed to increase reporting quality, completeness, and consistency. Currently, the Sector Standards cover the biggest-impact industries such as the oil and gas, coal, and the agriculture, aquaculture, and fishing sector. The standards list the topics that are likely to be material for most organisations in each sector. Hence, it aims to help companies at identifying relevant disclosures.

GRI Topic Standards

Each standard provides an overview of a particular topic and disclosure specific to it. Examples include standards on energy, emissions, diversity and equal opportunity, and anti-corruption. The topics are classified into three main categories: 200 economic aspects; 300 environmental aspects; and 400 social aspects (Table 8.1).

8.1.2 Integrated Reporting (IR) Framework

Historical Information About the IIRC

The International Integrated Reporting Council (IIRC) was a global coalition of regulators, investors, businesses, standard regulators, standard setters, accounting professionals, and non-governmental organisations (NGOs). Together, this coalition shared the vision that communicating value creation should be the next step in the development of corporate and sustainable reporting. Integrated reporting (IR) was therefore developed to respond to this need. The integrated report brings together material information about an organisation's strategy, governance, performance, and prospects in a way that reflects the business, social, and environmental context, providing a clear and concise representation of how an organisation creates and

Table 8.1 GRI Topic Standards

Topic	GRI Standards
Economic	GRI 202: Market Presence 2016
	GRI 203: Indirect Economic Impacts 2016
	GRI 204: Procurement Practices 2016
	GRI 205: Anti-corruption 2016
	GRI 206: Anti-competitive Behavior 2016
	GRI 207: Tax 2019
Environmental	GRI 301: Materials 2016
	GRI 302: Energy 2016
	GRI 303: Water and Effluents 2018
	GRI 304: Biodiversity 2016
	GRI 305: Emissions 2016
	GRI 306: Effluents and Waste 2016
	GRI 306: Waste 2020
	GRI 308: Supplier Environmental Assessment 2016
Social	GRI 401: Employment 2016
	GRI 402: Labor/Management Relations 2016
	GRI 403: Occupational Health and Safety 2018
	GRI 404: Training and Education 2016
	GRI 405: Diversity and Equal Opportunity 2016
	GRI 406: Non-discrimination 2016
	GRI 407: Freedom of Association and Collective Bargaining 2016
	GRI 408: Child Labor 2016
	GRI 409: Forced or Compulsory Labor 2016
	GRI 410: Security Practices 2016
	GRI 411: Rights of Indigenous Peoples 2016
	GRI 413: Local Communities 2016
	GRI 414: Supplier Social Assessment 2016
	GRI 415: Public Policy 2016
	GRI 416: Customer Health and Safety 2016
	GRI 417: Marketing and Labeling 2016
	GRI 418: Customer Privacy 2016

maintains value. An important aspect of the IR is to show connections between a company's financial performance and sustainability performance.

To ensure a global acceptance of the IR framework, the IIRC established a pilot programme allowing companies to become familiar with this new reporting concept and share their experience. An essential part of enabling global acceptance is establishing the concepts and principles upon which the final IR framework would be created. Therefore, the pilot programme played a key role in testing the robustness of the first trials.

The King III Governance Code, published in South Africa in September 2009 under the leadership of Mervyn King, required companies listed on the Johannesburg Stock Exchange (JSE) to submit an integrated report of their financial years beginning on or after 1 March 2010. Therefore, South African companies listed on the JSE were the first group of listed companies in the world to produce integrated reports, which replaced the old form of annual reports that focused primarily on financial information and the short-term horizon. An integrated report is a promising tool which allows stakeholders to better evaluate a company's performance and conduct an informed assessment of its ability to create and sustain value (Krzus & Eccles, 2010). Over 75 companies from 23 countries thus voluntarily enrolled in IR in the first year of the pilot programme.

Integrated Reporting Framework

The first version of the IR Framework was published in 2013. The current revised version was issued in 2021. The integrated report (IR) promotes a more cohesive and efficient approach to corporate reporting and equally aims to improve the quality of information available to financial capital providers which would lead to a more efficient allocation of capital.

The main objective of an integrated report is to explain to financial capital providers how an organisation creates and sustains value over time.

The IR Framework adopts a principles-based approach with the intention of providing a certain level of flexibility, recognising the different circumstances in which organisations can find themselves.

An integrated report aims to provide information about an organisation's resources and relationships. The latter are called types of "capitals". On the other hand, the report seeks to explain how the organisation interacts with the external environment and with capital to create value in the short, medium, and long term. Capitals are inventories of value that increase, decrease, and undergo transformations through a company's activities. The framework establishes six categories of capital:

- Financial capital
- Manufactured capital
- Intellectual capital
- Human capital
- Social and relationship capital
- Natural capital

A central point of the report is a description of the business model, which is an organisation's system of converting inputs through business activities into outputs and outcomes that create value over the short, medium, and long term (Fig. 8.3).

Integrated reporting is based on seven guiding principles that underpin the preparation and presentation of the IR:

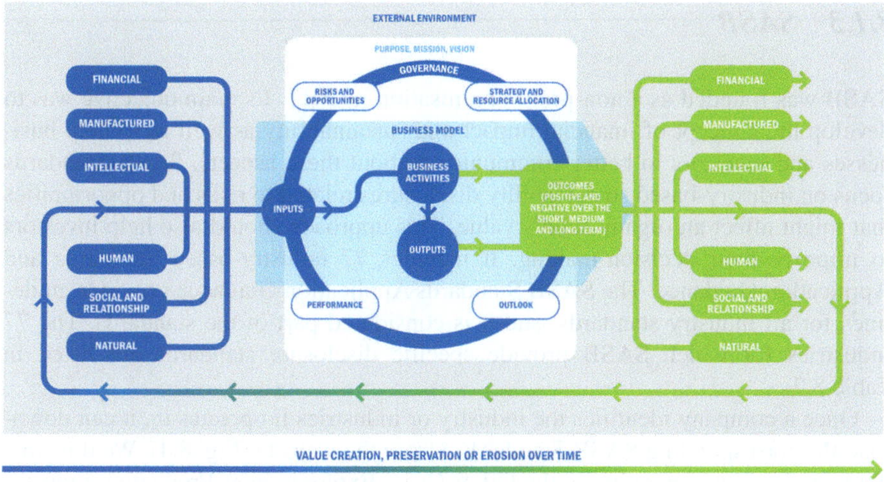

Fig. 8.3 Process of value creation, preservation, or erosion (IR, 2021)

- Strategic focus and future orientation
- Connectivity of information
- Stakeholder relationships
- Materiality
- Conciseness
- Reliability and completeness
- Consistency and comparability

Hence, these principles provide organisations with guidance on *how* to report.

In addition, an integrated report includes the following eight content elements to help organisations with *what* to report:

- Organisational overview and external environment
- Governance
- Business model
- Risks and opportunities
- Strategy and resource allocation
- Performance
- Outlook
- Basis of preparation and presentation

The concept of integrated reporting was created to meet twenty-first-century stakeholder information needs. Based on the above-mentioned reporting principles and concept, it should help different stakeholders including investors, employees, customers, suppliers, partners, local communities, legislators, regulators, and politicians, to make informed decisions.

The IIRC recently merged with SASB to create the Value Reporting Foundation, which has been operating under the umbrella of the IFRS Foundation since 2022.

8.1.3 SASB

SASB was founded as a non-profit organisation in 2011. Its main objective was to develop the concept of financial impact and sustainability as well as to help businesses and investors to better communicate about these aspects. SASB Standards focus on industry-based sustainability disclosures related to risks and opportunities that might affect an organisation's value. This approach should also help investors to improve their decision-making. It includes 77 industry-based Standards and Application Guidance. The SASB Standards Application Guidance provides guidelines for all industry standards, and it is considered part of the standards. The 77 industries for which SASB provide specific disclosure standards are listed in Table 8.2.

Once a company identifies the industry or industries it operates in, it can download the corresponding SASB Standard(s) from the website (Fig. 8.4). We illustrate below a company operating in the Oil & Gas—Exploration & Production industry from the Extractives & Materials Processing sector.

Figure 8.5 shows the initial page of the downloaded SASB Industry Standard related to Oil & Gas—Exploration & Production.

After opening the standard, the list of sustainability disclosure topics and metrics is displayed for a particular industry. In the case of the Oil & Gas—Exploration and Production industry, a company would be recommended to disclose the following topics: greenhouse gas emissions, air quality, water management, biodiversity impacts, etc. (Fig. 8.6).

Hence, the standard guides a company in relation to what it is expected to disclose regarding a particular industry, disclosure categories, acceptable methods of disclosure measurement, etc.

The SASB Standards are now an integral part of the ISSB Standards, and their continued use is expected to support companies on the path to successful ISSB implementation. Table 8.3 outlines how SASB contributes to IFRS Sustainability Disclosure Standards.

8.1.4 SDGs

Sustainable Development Goals (SDGs) were adopted in 2015 as part of Agenda 2030. Since then, an increasing number of companies and organisations around the world are disclosing how they are contributing to their achievement. This subchapter will shed some light on the scope of SDG reporting. Disclosure on the SDGs leverages the Ten Principles of the UN Global Compact as well as the GRI Standards because the latter are the mostly widely used globally. In August 2018, the UN

Table 8.2 The list of sectors and industries for SASB Standards (SASB, 2023a)

Sector	Industries
Consumer Goods	Apparel, Accessories & Footwear; Appliance Manufacturing; Building Products & Furnishings; Household & Personal Products; Toys & Sporting Goods; Multiline and Specialty Retailers & Distributors; E-commerce
Extractives & Minerals Processing	Coal Operations; Construction Materials; Iron & Steel Producers; Metals & Mining; Oil & Gas—Exploration & Production; Oil & Gas—Midstream; Oil & Gas—Refining & Marketing; Oil & Gas - Services
Financials	Asset Management & Custody Activities; Investment Banking & Brokerage; Security & Commodity Exchanges; Commercial Banks; Consumer Finance; Mortgage Finance; Insurance
Food & Beverage	Agricultural Products; Meat, Poultry & Dairy; Processed Foods; Alcoholic Beverages; Non-Alcoholic Beverages; Food Retailers & Distributors; Restaurants; Tobacco
Health Care	Biotechnology & Pharmaceuticals; Drug Retailers; Health Care Delivery; Health Care Distributors; Managed Care; Medical Equipment & Supplies
Infrastructure	Electric Utilities & Power Generators; Gas Utilities & Distributors; Water Utilities & Services; Engineering & Construction Services; Home Builders; Real Estate; Real Estate Services; Waste Management
Renewable Resources & Alternative Energy	Biofuels; Fuel Cells & Industrial Batteries; Solar Technology & Project Developers; Wind Technology & Project Developers; Forestry Management; Pulp & Paper Products
Resource Transformation	Aerospace & Defense; Containers & Packaging; Electrical & Electronic Equipment; Industrial Machinery & Goods; Chemicals
Services	Advertising & Marketing; Media & Entertainment; Casinos & Gaming; Hotels & Lodging; Leisure Facilities; Education; Professional & Commercial Services
Technology & Communications	Electronic Manufacturing Services & Original Design Manufacturing; Hardware; Software & IT Services; Internet Media & Services; Semiconductors; Telecommunication Services
Transportation	Airlines; Air Freight & Logistics; Automobiles; Auto Parts; Car Rental & Leasing; Cruise Lines; Marine Transportation; Rail Transportation; Road Transportation

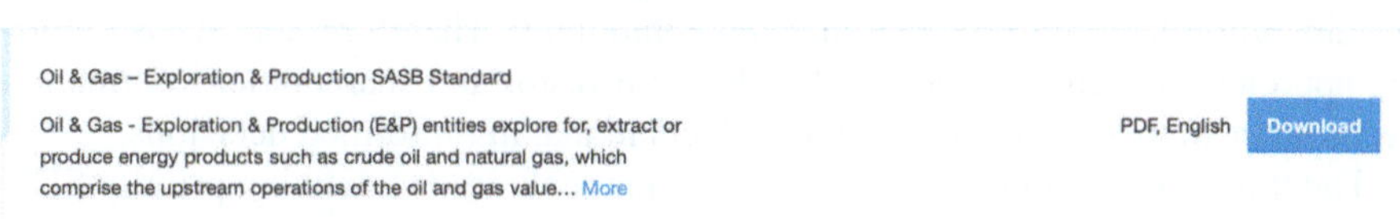

Fig. 8.4 Downloading the SASB Standard (SASB, 2023a)

SASB STANDARDS
Now part of IFRS Foundation

Oil & Gas – Exploration & Production
Sustainability Accounting Standard

EXTRACTIVES & MINERALS PROCESSING SECTOR
Sustainable Industry Classification System® (SICS®) EM-EP
Under Stewardship of the International Sustainability Standards Board

INDUSTRY STANDARD | VERSION 2023-06

Fig. 8.5 SASB Standards Oil & Gas (SASB, 2023b)

Sustainability Disclosure Topics & Metrics 6
Greenhouse Gas Emissions 9
Air Quality 15
Water Management 17
Biodiversity Impacts 21
Security, Human Rights & Rights of Indigenous Peoples 26
Community Relations 30
Workforce Health & Safety 34
Reserves Valuation & Capital Expenditures 37
Business Ethics & Transparency 42
Management of the Legal & Regulatory Environment 45
Critical Incident Risk Management 47

Fig. 8.6 Sustainability Disclosure Topics and Metrics for Oil & Gas industry (SASB, 2023b)

Global Compact together with GRI and the PWC's technical support issued a *Practical Guide Integrating the SDGs into Corporate Reporting*. This practical guide outlines a three-step process to embed the SDGs not only into a company's existing reporting processes but also into its business processes. The Principles for Responsible Investment (PRI) are also incorporated to help increase the value of sustainability disclosure for the financial community. Therefore, the SDGs reporting approach is designed not only to address investor and shareholder information needs but also to assist different stakeholders in making informed decisions.

The three-step process of the practical guide includes (1) defining priority SDG targets (understanding the SDGs and their targets, conducting prioritisation of SDG targets, defining company SDG-related report content); (2) measurement and analysis (setting business objectives, selecting appropriate disclosures, collecting and analysing data); and (3) reporting, integrating, and implementing change (considering general features of good reporting practices, considering user information needs, reporting and change implementation) (UNGC, 2018).

Table 8.3 SASB contribution to IFRS Sustainability Disclosure Standards (SASB, 2023c)

Disclosure standards	Description
IFRS S1	The SASB standards help identify sustainability-related risks and opportunities and disclose related information beyond climate
IFRS S2	The SASB standards includes accompanying guidance derived from the climate-related topics and metrics, which have been enhanced to improve international applicability

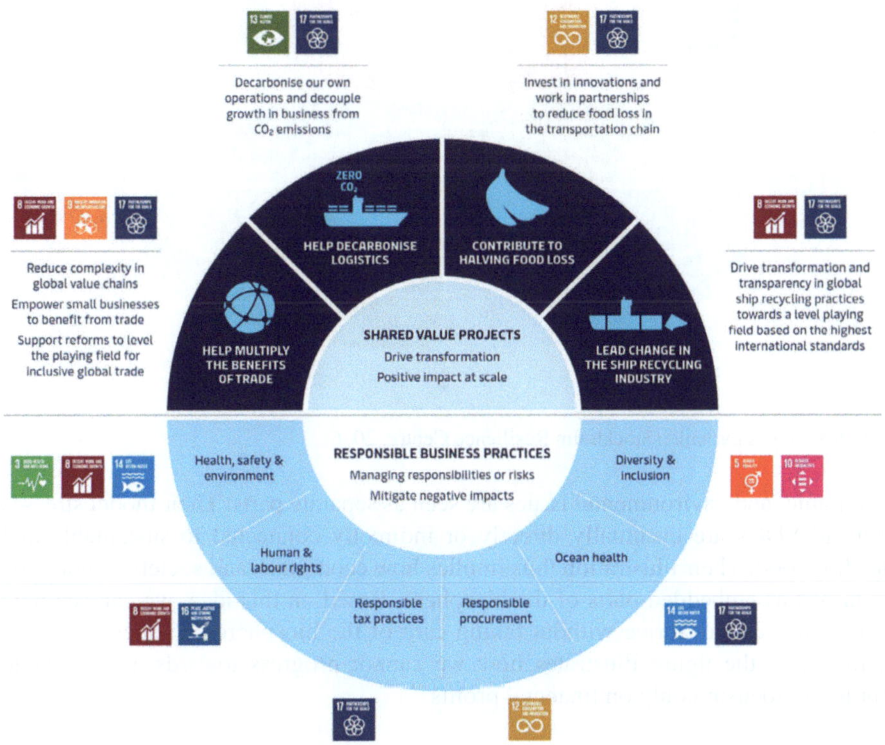

Fig. 8.7 Linking objectives and SDGs of Maersk (Maersk, 2018)

Figure 8.7 outlines the current SDGs corporate reporting practices. It depicts how the Maersk company links its sustainability objectives to SDGs.

The SDGs Wedding Cake

The discussion of particular SDGs is beyond the scope of this book, but we do present an alternative way of viewing the SDGs in Fig. 8.8. This approach is provided by the Stockholm Resilience Centre of Stockholm University. Their vision provides a perspective that differs from the current sectorial approach where social,

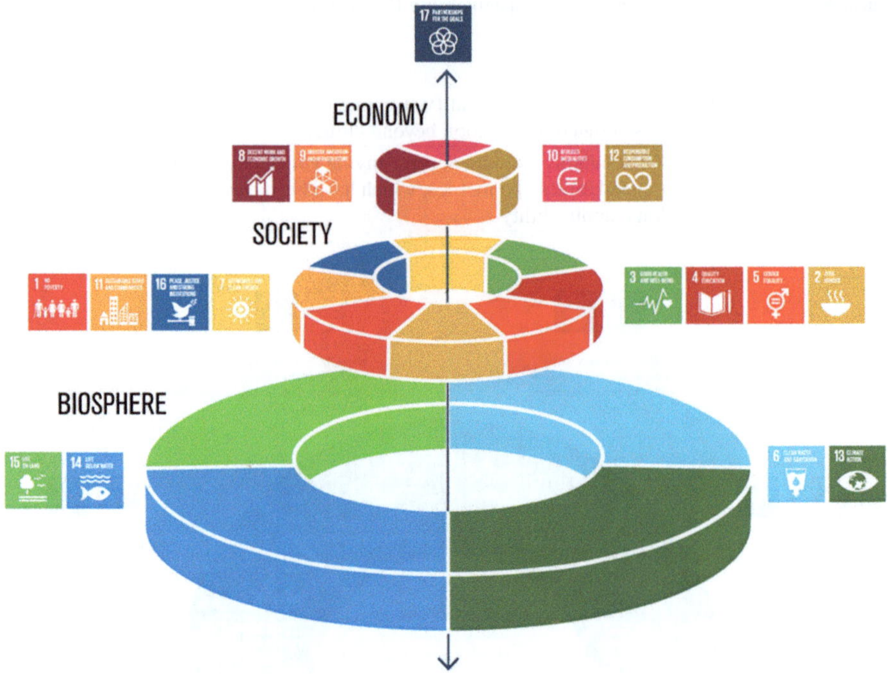

Fig. 8.8 SDGs pyramid (Stockholm Resilience Centre, 2016)

economic, and environmental issues are seen as separate parts. Their model stresses that all SDGs are eventually directly or indirectly connected to sustainable and healthy food. Their illustration thus implies how economies and societies should be regarded as embedded parts of the biosphere. Based on this idea, we can see how economies cannot thrive without taking care of the biosphere and society. At the same time, the figure illustrates how we cannot progress towards a sustainable future by focusing only on financial profits.

8.1.5 Importance of Alignment in ESG/ Sustainability Reporting

The GRI Standards are still the most dominant globally, and the SASB is currently the leading reporting standard in the USA, Canada, and Brazil. However, we have recently been witnessing the growing adoption of country stock exchange guidelines, particularly in African countries and in Malaysia as well as India. This multiplication of diverse sustainability reporting standards makes it rather challenging to compare companies and countries. Yet, if we wish to address the major Agenda 2030 sustainability issues collectively, it is essential that we all "speak the same language". In other words, it is becoming ever more important that sustainability reporting frameworks converge. Today's major alignment initiatives are ISSB and CSRD, which we discuss briefly below.

ISSB

In 2021, SASB and the International Integrated Reporting Council (IIRC) merged to form the Value Reporting Foundation (VRF). In November 2021, the International Sustainability Standards Board (ISSB) was established by the IFRS Foundation. The aim of the ISSB is to develop standards for a global baseline of sustainability disclosures. In August 2022, the Value Reporting Foundation (a merger of the SASB and the IIRC) was consolidated into the IFRS Foundation whose main objective is to drive connectivity between sustainability disclosure and financial statements (VRF, 2023). Thus, integrated reporting concepts and principles are embedded into the ISSB Standards. Similarly, ISSB is also in line with the SASB approach. In June 2023, the ISSB issued its inaugural global IFRS Sustainability Disclosure Standards, IFRS S1 and IFRS S2 (Table 8.4). The IFRS Foundation's International Accounting Standards Board (IASB) and the International Sustainability Standards Board (ISSB) collaborate on integrating the IR framework and SASB Standards into their standard setting projects (IR, 2022). This initiative does thus not operate in isolation; it builds instead on the work of well-established, existing reporting initiatives such as the Integrated Reporting Framework, the SASB Standards, but also the Climate Disclosure Standards Board (CDSB) and the Task Force for Climate-related Financial Disclosures (TCFD). In addition, it has the support of the G7, the G20, the

Table 8.4 IFRS Sustainability Disclosure Standards

IFRS sustainability disclosure standards	Title	Objective	Effective day
IFRS S1	General Requirements for Disclosure of Sustainability-related Financial Information	The objective of IFRS S1 is to require an entity to disclose information about its sustainability-related risks and opportunities that is useful to users of general purpose financial reports in making decisions relating to providing resources to the entity	Annual reporting period beginning on or after 1 January 2024
IFRS S2	Climate-related Disclosures	The objective of IFRS S2 is to require an entity to disclose information about its climate-related risks and opportunities that is useful to users of general purpose financial reports in making decisions relating to providing resources to the entity	Annual reporting period beginning on or after 1 January 2024

Author's elaboration based on IFRS (2023a, b)

International Organization of Securities Commissions (IOSCO), Finance Ministers, and Central Bank Governors from more than 40 jurisdictions (IFRS, 2023a, b).

The IASB and ISSB have committed to a long-term collaboration regarding sustainability disclosure standards, incorporating principles and concepts from the Integrated Reporting Framework together with well-developed, industry-specific SASB Standards, assembling them into a cohesive whole. This effort promises a holistic and comprehensive sustainability disclosure approach.

Corporate Sustainability Reporting Directive (CSRD) and European Sustainability Reporting Standards (ESRS)

In April 2021, the European Commission (EC) issued the CSRD proposal, which will require large European companies to report on their social and environmental performance. A year later, in April 2022, EFRAG issued the first draft of the European Sustainability Reporting Standards (ESRS).

The EU updated the first regulation on non-financial disclosure on 22 October 2014, when the EU Council adopted a Non-financial Reporting Directive (NFRD). This directive amended Directive 2013/34/EU on annual financial statements, consolidated financial statements, and related reports of certain types of companies (EUR-Lex, 2014). Its main objective was to encourage further transparency by establishing minimal requirements regarding the extent of the non-financial information that should be made available to the public. However, it still offered a rather high degree of freedom regarding the extent and content of the sustainability report. Thus, companies were free to choose a reporting framework such as GRI, UN, OECD, ISO 26000, SA 8000, etc. or national standards.

On 5 January 2023, the Corporate Sustainability Reporting Directive (CSRD) entered into force. This new directive revises and extends the scope of the sustainability reporting requirements introduced by the NFRD (EUR-Lex, 2022). At the same time, it modernises and strengthens the rules about the social and environmental information that companies must report. Moreover, a broader set of large companies, as well as listed SMEs, will now be required to report on sustainability—approximately 50,000 companies in total. The first companies will have to apply the new rules for the first time in financial year 2024, for reports published in 2025. Companies subject to the CSRD will have to report according to European Sustainability Reporting Standards (ESRS), developed by the EFRAG. The CSRD also makes it mandatory for companies to undergo an audit of their reported sustainability information. This should increase the credibility of the reported information. In addition, new regulation encourages the digitalisation of sustainability information through an XHTML format in accordance with the European Single Electronic Format (ESEF) intended to boost comparability.

The new ESRS standards were built considering the existing standards such as the Global Reporting Initiative, the Sustainability Accounting Standards Board, the International Integrated Reporting Council, the International Accounting Standards Board, the Task Force on Climate-Related Financial Disclosures, the Carbon

Disclosure Standards Board, and the CDP (Carbon Disclosure Project). The intention is to ensure an alignment with other EU initiatives on sustainable finance—in particular, the Sustainable Finance Disclosure Regulation (SFDR).

In addition, to avoid unnecessary sustainability standard fragmentation, the European standards also aim to contribute to the process of global convergence of sustainability reporting standards by supporting the efforts of the International Sustainability Standards Board (ISSB).

In July 2023, the EC published the final text of the first set of *12 ESRSs*. One of the key features of the final set of standards is the double materiality principle, which requires companies to include information that is material from either a financial perspective or an impact perspective. In addition, reporting from across the value chain should be included.

Figure 8.9 outlines the current scope of the ESRS, which includes topic-specific disclosure requirements related to governance, strategy, impact, risk, and opportunity management. It also establishes metrics and explains how to report targets relating to each topic.

According to the ESRS, companies need to disclose *four reporting areas* such as governance; strategy; impact, risk, and opportunity management; and metrics and targets of selected sustainability topics. There are *three reporting layers:* (1) sector agnostic disclosures, applicable to all companies; (2) sector-specific disclosures, currently under development but that will be applicable to specific sectors; (3) company-specific disclosures, which would refer to additional disclosure on material topics specific to the company. The sustainability report should cover *three main topics:* the environmental domain (climate change, pollution, water and marine resources, biodiversity and ecosystems, resource use and circular economy); the

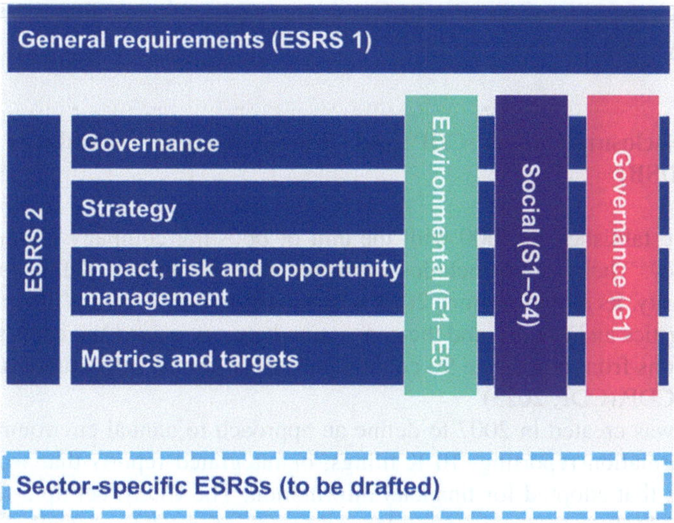

Fig. 8.9 Structure of the ESRS (KPMG, 2023)

social domain (their own workforce, workers in the value chain, affected communities, consumers and end users); and governance (business conduct) (KPMG, 2023).

8.1.6 Other Relevant Sustainability Disclosure Initiatives

Task Force on Climate-Related Financial Disclosures (TCFD)

TCFD was created by the Financial Stability Board in 2015 to develop corporate disclosure recommendations to support actors in financial markets such as investors, lenders, etc. in appropriately assessing a set of risks related to climate change for efficient capital-allocation decisions. In 2017, the TCFD issued climate-related financial disclosure recommendations, which are structured around four thematic areas—governance, strategy, risk management, as well as metrics and targets—related to 11 recommended topics that should help investors and other stakeholders to operate on financial markets and to understand how companies approach climate-related risks and opportunities (TCFD, 2023).

Principles for Responsible Investments (PRI)

PRI is an independent initiative supported by the United Nations (UN) and the world's leading proponent of responsible investment. PRI aims to encourage investors to invest in a responsible way by incorporating ESG factors into their investment and ownership decisions. Nowadays, it is a collaborating partner in SDG reporting together with the UN Global Compact and GRI Standards. The PRI's role is to help increase the value of corporate sustainability disclosure for the financial community (PRI, 2023).

Carbon Disclosure Project (CDP) and Climate Disclosure Standards Board (CDSB)

CDP was established in 2000 with the aim of boosting company climate impact transparency. Their scope includes environmental disclosure, deforestation, and water security. Its ambition since 2021 is to expand to new domains linked to biodiversity, plastics and oceans, and the connection between nature and earth's systems. Organisations from more than 90 countries disclose annual information in compliance with CDP (CDP, 2023).

CDSB was created in 2007 to define an approach to annual environmental and social information reporting, 10-K filings, or integrated reports that would be as rigorous as that adopted for financial information. The CDSB set up a foundation for the TCFD. It covers environmental and social information as well as technical guidance on climate, water, and biodiversity reporting. CDSB and CDP now work together to improve climate disclosure transparency (CDSB, 2022).

8.2 ESG Communication Strategies

ESG reporting, an essential tool for corporate transparency regarding its environmental and social accountability, has evolved beyond traditional annual reports and standalone sustainability or integrated reports. Nowadays, companies increasingly leverage alternative means of ESG communication such as social networks and social media, due to their interactive nature and ability to foster stakeholder engagement. Social networks represent dynamic and interactive platforms that offer numerous advantages for companies seeking to enhance their ESG communication strategies.

Social media such as Facebook, Twitter, LinkedIn, YouTube, Instagram, and others provide companies with the possibility to connect with a wide array of stakeholders directly and in real-time, including with investors, customers, employees, NGOs, etc. In addition, the interactive nature of these platforms allows for rapid dissemination and also facilitates two-way communication, which means that companies can share ESG-related content, receive immediate feedback, and engage with stakeholders.

Using social networks/media for ESG disclosure is not new. It became ever more legitimate to resort to SNs for corporate disclosure as stakeholders expected greater transparency and updated information on a company's ESG performance and impact. The effect of SNs has grown with the rising number of users and their daily presence on these media. In addition, stakeholders increasingly prefer concise and visually engaging disclosure over lengthy, traditional corporate reports.

Accounting scholars also claim that text disclosure is no longer adequate to effectively communicate ESG information. They stress the important role of visual imagery, such as pictures and videos to deliver the message more efficiently. Social networks offer an opportunity to deliver messages in various multimedia formats, including videos, pictures, infographics, etc.

From a psychological viewpoint, video is a particularly powerful format due to a combination of sight, sound, motion, and emotion reaching our senses. In general, ESG disclosure on these platforms represent an interesting tool to better engage stakeholders: indeed, compared to text disclosure in corporate reports, they offer a rich combination of added layers of media formats (photos, videos) together with social network-enabled user interactions. The increasing popularity of social networks offers companies a new means to build relationships with their stakeholders and can also lead to improving corporate reputation (Eberle et al., 2013).

The main reasons why companies are turning to social networks as an alternative means of ESG reporting are:

- Information shared on these platforms can reach a wide audience quickly, which increases the *visibility* and impact of a company's ESG initiatives, potentially influencing public perception and investor sentiment.
- Social networks enable answering stakeholder inquiries promptly, which increases responsiveness and helps build *trust and credibility*.
- Direct, real-time communication increases *transparency*.
- Two-way communication helps build better *stakeholder engagement*.

Nevertheless, when considering social networks and media for ESG communication, companies should establish a clear communication strategy, which should include aspects such as:

- *Mission and Vision*

- Why use these alternative tools for ESG disclosure and communication? How will they be integrated into the organisation's communication strategy?
- *Target Audience*
- Who is the target audience? This is a very important aspect to address as demographics differ from one social platform to another.
- *Selection of Social Platforms*
- Establish selection criteria such as utility for the organisation; where the users are; level of interaction, measurement of results; etc.
- *Available Human Resources*
- Who will manage them? What resources do companies have for this purpose? Promotion of new professional competencies; new professional profiles (e.g., community manager or social media strategist).
- *Operational Implementation Plan*
- Specific objectives to be developed; content generation tools; risk management.
- *Measurement and Evaluation*

- Attention to key variables; indicators, measurement instruments, and monitoring tools.

Before closing, it is important to stress that using alternative platforms for ESG reporting and communication requires careful management to ensure the accuracy and credibility of such disclosure. Therefore, companies should ensure that disseminated information aligns with real ESG performance and that these platforms are not used only to appeal to different stakeholders. In addition, companies must be ready to face criticism from the general public.

On the other hand, in practice, we can sometimes observe that companies take advantage of these interactive and engaging (therefore also persuasive) platforms for unethical purposes such as greenwashing or bluewashing, which will be explained in the next section.

8.3 Ethical Considerations in ESG Reporting and Transparency: Greenwashing and Bluewashing

Greenwashing and bluewashing are unethical practices in the context of ESG reporting. Both are based on giving misleading information or exaggerating an organisation's commitment to environmental (greenwashing) and social (bluewashing) responsibility. These deceptive practices emerged as ESG criteria played a growing role in investment decisions and stakeholder engagement. They are based on projecting a more favourable image than warranted by a company's actual ESG performance, potentially triggering reputation benefits and eventually financial gains.

Greenwashing is associated with environmental aspects of ESG disclosure. One example could be when a company portrays itself as more environmentally friendly than it truly is through misleading marketing or statements that suggest that it is strongly committed to environmental responsibility but with no substantiated action. This approach can mislead stakeholders and investors into believing that the company adheres to strict, environmentally-friendly practices when, in reality, it may fall short of those claims.

One illustration is the Volkswagen (VW) Emissions Scandal in 2015, when this leading automobile manufacturer marketed its diesel vehicles as low-emission, despite installing software that manipulated emissions during regulatory testing. The company thus projected an environmentally friendly image through advanced sustainability disclosure, but in reality, it was using a deceptive strategy to manipulate both consumers and regulators. The scandal caused significant harm to its reputation and had important financial repercussions.

Bluewashing is linked to the social and governance aspects of ESG. Companies present themselves as socially responsible or ethically governed, although their practices may not be aligned with such claims. It can involve superficial CSR initiatives or overstating efforts to project good and ethical governance.

The Rana Plaza factory collapse in Bangladesh in 2013 is an illustration of bluewashing practices. A number of global fashion brands—including Primark, Mango, and Benetton—were implicated in the tragedy, in which over a thousand workers were killed and thousands injured. Prior to the collapse, the brands involved had presented themselves as socially responsible and committed to ethical supply chain management. However, a supplier of their garments failed to ensure the safety and well-being of its factory workers. This accident increased awareness regarding bluewashing practices and the gap between glowing claims and actual social responsibility.

Within the realm of ESG reporting, greenwashing and bluewashing are unethical practices designed to project a misleading impression of a company's commitment to environmental and social responsibility or governance integrity. It is important to raise awareness about this issue, which can occur more easily in an unregulated area. Indeed, over the last 30 years, ESG reporting has mostly been voluntary and in some countries it still is. Voluntary and non-audited statements in this matter are a fertile soil for exaggerated estimations and inaccurate, appealing projections. Therefore, current regulatory initiatives such as Eur Lex (2022) which also require an audit of sustainability information, among other elements, are of utmost importance to combat these kinds of practices and increase trust in ESG reporting. Indeed, they ensure genuine adherence to ESG and sustainability principles, preventing ESG reporting from becoming a mere marketing tool.

8.4 Corporate Digital Responsibility (CDR) and ESGD Reporting

8.4.1 Rapid Evolution of New Technologies and Raising Ethical Concerns

Over the last decade we have witnessed how new technologies such as robotics, automation, artificial intelligence, etc. have increased not only production process efficiency but also that of other domains. The latter include marketing, where they contribute to creating customer profiles, or HR, where they assist recruitment processes, among others. Nevertheless, such processes come with certain ethical issues such as data protection, cybersecurity, safety, or human rights. Automated decision-making (ADM) algorithms raise particular concern as biased or erroneous ADMs represent a real risk when no human supervision is applied. And the negative implications can be huge. Therefore, it is important to be aware of the potential dangers of non-human-supervised AI and automation. One real-life example is Amazon's biased algorithms which led to hiring more men than women. While AI and automation certainly improve efficiency, they also trigger ethical problems such as discrimination, as in the above-mentioned example.

Over the last few years, based on an omnipresent philosophy of progress, we have observed much excitement around AI and superintelligence, particularly after the launch of generative AI such as ChatGPT. We must be cautious, however, and adopt a proactive approach to "what can go wrong" in order to anticipate and mitigate catastrophic risks: only by understanding AI risks and potential dangers will we be able to mitigate them. Therefore, efforts to ensure that AI is safely developed and deployed should be made collectively and proactively.

Regarding "what can go wrong", it is worth reflecting on the competition that tech giants are currently engaged in. Competition among companies is normally a necessary, healthy phenomenon in an economy. But in the race towards AI solutions, it can have highly negative consequences as the pressure to develop quickly without proper risk management can lead to sacrificing safety and collective well-being for the sake of efficiency, profits, and the benefit of only a few.

In response to AI's rapid evolution, the European Commission proposed an AI regulatory framework in April 2021. The EU Parliament voted its position in June

2023 and lawmakers are now negotiating to finalise the new legislation. The AI Act will adopt a risk-based approach, dividing AI systems into categories such as unacceptable risk (these AI tools will be banned, an example could be an AI system for social scoring); high risk (AI used in domains considered risky such as product safety, aviation, medical devices, etc.); limited risk (e.g., generative AI such as ChatGPT or chatbots); and low and minimal risks. Based on this approach, AI systems in the high-risk category would require compliance with certain requirements and higher transparency. AI systems in the limited-risk category would require certain transparency, e.g., users should be notified that they are dealing with a machine and not a human or that a content has been generated by AI.

As we can observe, a significant amount of current risks to society stem from AI. Nevertheless, surprisingly, the latest European Sustainability Reporting Standards issued by EFRAG in 2023 do not reflect this dimension in the disclosure requirements.

While the CSR philosophy and movement aimed to address environmental and social challenges (such as gender inequality, supply chain threats, etc.), today, we are facing new threats that should be addressed accordingly. Indeed, new threats imply new responsibilities for companies, such as corporate digital responsibility (CDR).

CDR is a relatively new phenomenon, but some large companies provide voluntary disclosure in this domain. The scope of CDR, however, is not very clear: it spans from online shopping, to waste and carbon footprint reduction, to digital literacy, and to combatting cyberbullying.

8.4.2 CDR and Sustainability

In addition to increasing efficiency, digitalisation has a strong potential to contribute to sustainable development. Yet, digital transformation is currently failing to create sustainability due to a lack of responsibility and trust. So why is it necessary to guarantee sustainable development? On the one hand, digital transformation in the form of AI, automation, etc. can potentially improve people's lives and the efficiency of economic processes. On the other, it can help to build a trustworthy digital sphere, the basis for sustainable development. A company engaging in the world of new technologies such as algorithmic automation, artificial intelligence, Internet of Things, blockchain, etc. should critically reflect on different ethical criteria to avoid compromising democratic rights and values (Mueller, 2022).

Therefore, to move towards sustainable development, ESG should be given an additional "digital" dimension (Bednárová, 2022). Nevertheless, while digitalisation can accelerate the development of efficient social and economic processes, the ability to identify and mitigate ethical concerns related to digital technologies will represent a key skill in sustainable development.

SDGs have been developed to advance towards sustainable development and they represent an agenda for a more sustainable planet. Considering the importance of digitalisation and its role in sustainable development, there is a need to focus more closely on this area as well.

The recognition that technology implementation comes with certain responsibilities leads to a relatively new concept of corporate digital responsibility (CDR). CDR emerged in the literature recently, but the subject is already gaining momentum in both research and practice. Academic definitions of CDR are outlined below. As we can observe, for some academics, there is a clear connection between sustainability, ESG factors and CDR.

> … a voluntary corporate orientation to ensure a responsible use of digital technologies. Weißenberger and Marrocco (2022)

> … a set of shared values and norms guiding an organisation's operations with respect to the creation and operation of digital technology and data. Lobschat et al. (2021)

> … an extension of a firm's responsibilities which takes into account the ethical opportunities and challenges of digitalisation. Herden et al. (2021)

> … a set of practices and behaviors that help an organisation use data and digital technologies in a way that is socially, economically, and environmentally responsible. Wade (2020)

8.4.3 CDR Disclosure Initiatives

The degree of CDR awareness varies according to the country, but today, Germany can be considered as a leader in CDR initiatives. Germany's Federal Ministry of Justice and Consumer Protection introduced the CDR Initiative back in 2018. This new initiative is supported by both the government and a number of renowned listed companies. And the initiative has produced the CDR Code which establishes objectives in five fields of action: data handling, education, climate and resource protection, employee involvement, and inclusion. The main aim is to ensure that digital responsibility becomes an integrated part of day-to-day business decisions (BMJV, 2018). At the same time, this initiative can be considered as an initial effort to formalise CDR corporate disclosure.

Similarly, the Spanish Accounting and Administration Association (AECA) has recently been working on a CDR disclosure reporting framework. They defined the CDR as follows: *CDR consists of a corporate orientation towards a legal and ethical use of data and digital technologies that protects people's rights regarding the data and algorithmic decisions that affect them, ensuring trust in the security, usefulness, and efficiency of the technology* (AECA, 2022).

8.4.4 *How CDR Can Be Implemented and Enforced in Corporations*

Some companies recognise the increasing importance of CDR and have created departments or offices that address ethical issues related to new technologies, e.g. Merck's Digital Ethics Advisory Panel.

Implementing and enforcing CDR in corporations requires a complex approach. The process can consist of, but it is not limited to, the following steps:

- Changes in an organisation's hierarchical and procedural structures (creation of a corresponding office or naming an officer as an authority to police CDR)
- Incorporating CDR into a company's CSR and recognising its implications for ESG dimensions
- Development of a CDR framework
- CDR disclosure and transparency to provide accountability and ensure trust in digital products and services

8.4.5 *CDR Implications on ESG*

Innovation and digitalisation are complementary elements that are necessary for sustainability. Due to the increasing digitalisation of activities and processes, corporate responsibility has been currently expanded from ESG to include a digital dimension, resulting in an ESGD approach. Such an approach lays the foundations for a new perspective on sustainability, where environmental, social, and governance dimensions are affected by digitalisation.

8.5 Summary

In this chapter, we embarked on a comprehensive exploration of ESG reporting and communication and have gained essential insights useful to navigate the complex landscape of sustainability and ESG disclosure. The focal point was the introduction of global ESG reporting frameworks and standards, with a primary emphasis on the widely implemented Global Reporting Initiative (GRI) Standards, a cornerstone for sustainability reporting worldwide. Additionally, we stressed the key role of Integrated Reporting (IR), in line with International Financial Reporting Standards (IFRS), alongside the exploration of other influential standards such as SASB and the Sustainable Development Goals (SDGs).

We also outlined alignment initiatives, focusing on the ISSB and European Sustainability Reporting Standards (ESRS) introduced with the CSRD regulation. In addition, we referred to disclosure initiatives such as the TCFD, PRI, and CDSB, which help to build an understanding of the dynamic and evolving nature of ESG reporting.

Nevertheless, we have not merely discussed standards and reporting frameworks. We have also provided insights into the intricacies of effective communication strategies for conveying ESG performance and impact. Ethical considerations such as greenwashing and bluewashing were addressed, raising awareness of the ethical dimensions surrounding ESG reporting and communication.

A forward-looking perspective was embraced with the introduction of a novel concept: corporate digital responsibility (CDR) within the realm of sustainability and corporate social responsibility (CSR). This was accompanied by an exploration of a potential paradigm shift of sustainability/ESG reporting from ESG to ESGD, which tackles the challenges posed by cutting-edge technologies such as artificial intelligence and automated decision-making systems.

This chapter equips students not only with an understanding of the world's major ESG reporting standards and frameworks but also with the analytical and ethical skills needed to navigate the evolving landscape of sustainability reporting and communication in the global business arena. The knowledge acquired extends beyond traditional reporting, encompassing the dynamic intersections of ESG communication via social media, and the evolving field of CDR, which covers technology, ethics, and corporate responsibility.

8.6 Discussion Questions

The following questions encourage students to critically engage with the content and reflect on the presented material. They are designed to promote a deeper understanding of the topic, stimulate thinking, and encourage further exploration.

1. Select a listed company from your country and identify its ESG reporting strategy by answering the following questions: (1) Which standards does the company have to comply with; (2) where does the company report on ESG performance? Is it within the management report, in a section of the annual report, in a separate sustainability report, in an integrated report, or elsewhere?
2. Which company has a well-developed ESG communication strategy on social media? Explain why.
3. Can you identify other examples of greenwashing and bluewashing tactics?
4. Can you identify a company which should be more transparent about its CDR? Explain why.

References

AECA. (2022). *Responsabilidad Digital Corporativa*. Opinión Emitida.

Bednárová, M. (2022). *Digital disruption and corporate digital responsibility*. Logos Verlag.

BMJV. (2018). Federal Ministry of justice and consumer protection. *CDR initiative*. https://cdr-initiative.de/en/initiative

Bowen, H. R. (1953). *Social responsibilities of the businessman*. Harper.

Carroll, A. (1991). Corporate social responsibility: Evolution of a definitional construct. *Business & Society, 38*(3), 268–295. https://doi.org/10.1177/000765039903800303

CDP. (2023). *CDP disclosure insight action*. https://www.cdp.net/en/info/about-us

CDSB. (2022). *Climate disclosure standards board*. https://www.cdsb.net/cdp

Eberle, D., Berens, G., Li, T. (2013). The impact of interactive corporate social responsibility communication on corporate reputation. *Journal of Business Ethics, 118*, 731–746.

EUR Lex. (2022). *CSRD Directive*. https://eur-lex.europa.eu/legal-content/EN/TXT/?uri=CELEX:32022L2464

EUR-Lex (2014). Directive 2014/95/EU. https://eur-lex.europa.eu/eli/dir/2014/95/oj/eng

Freeman, E. (1984). *Strategic management: A stakeholder approach*. Pitman Publishing.

GRI. (2021). *A short introduction to the GRI standards*. https://www.globalreporting.org/media/wtaf14tw/a-short-introduction-to-the-gri-standards.pdf

Herden, C. J., Alliu, E., Cakici, A., Cormier, T., Deguelle, C., Gambhir, S., Griffiths, C., Gupta, S., Kamani, S. R., Kiratli, Y.-S., Kispataki, M., Lange, G., Moles de Matos, L., Tripero Moreno, L., Betancourt Nunez, H. A., Pilla, V., Raj, B., Roe, J., Skoda, M., & Edinger-Schons, L. M. (2021). Corporate digital responsibility. *Sustainability Management Forum, 29*(1), 13–29. https://doi.org/10.1007/s00550-020-00509-x

IFRS. (2023a). *About the international sustainability standards board*. https://www.ifrs.org/groups/international-sustainability-standards-board/

IFRS. (2023b). *IFRS S1 general requirements for disclosure of sustainability-related financial information*. https://www.ifrs.org/issued-standards/ifrs-sustainability-standards-navigator/ifrs-s1-general-requirements/

IR. (2021). *Integrated reporting framework*. http://www.integratedreporting.org/wp-content/uploads/2022/08/IntegratedReportingFramework_081922.pdf

IR. (2022). *Integrated Reporting—articulating a future path*. https://www.integratedreporting.org/news/integrated-reporting-articulating-a-future-path/

KPMG. (2022). *Big shifts, small steps. Survey of Sustainability Reporting 2022*. https://assets.kpmg.com/content/dam/kpmg/se/pdf/komm/2022/Global-Survey-of-Sustainability-Reporting-2022.pdf

KPMG. (2023). *Get ready for European Sustainability Reporting Standards. Understanding the first set of ESRSs*. https://assets.kpmg.com/content/dam/kpmg/xx/pdf/2022/07/talkbook-get-ready-for-esrs.pdf

Krzus, M. P., & Eccles, R. G. (2010). *One report. Integrated reporting for sustainable strategy*. Wiley & Sons.

Lobschat, L., Mueller, B., Eggers, F., Brandimarte, L., Diefenbach, S., Kroschke, M., & Wirtz, J. (2021). Corporate digital responsibility. *Journal of Business Research, 122*, 875–888. https://doi.org/10.1016/j.jbusres.2019.10.006

Maersk. (2018). *Sustainability report*. https://www.maersk.com/sustainability/reports-and-resources

Mueller, B. (2022). Corporate digital responsibility. *Business Information Systems Engineering, 64*(5), 689–700. https://doi.org/10.1007/s12599-022-00760-0

PRI. (2023). *Principles for Responsible Investment*. https://www.unpri.org/about-us/about-the-pri

Roca, L. C., & Searcy, C. (2012). An analysis of indicators disclosed in corporate sustainability reports. *Journal of Cleaner Production, 20*, 103–118.

SASB. (2023a). *SASB standards*. https://sasb.org/standards/download/

SASB. (2023b). *Industry standards oil & gas*. https://d3flraxduht3gu.cloudfront.net/latest_ standards/oil-and-gas-exploration-and-production-standard_en-gb.pdf
SASB. (2023c). *The evolution of the SASB Standards*. https://sasb.org/blog/the-evolution-of-the-sasb-standards-a-conversation-with-jeff-hales/
Stockholm Resilience Centre. (2016). https://www.stockholmresilience.org/images/18.36c25848153d54bdba33ec9b/1465905797608/sdgs-food-azote.jpg
Suchman, M. C. (1995). Managing legitimacy: Strategic and institutional approaches. *The Academy of Management Review, 20*(3), 571–610.
TCFD. (2023). *Task force on climate-related financial disclosures*. History. https://www.fsb-tcfd.org/about/#history
The Value Reporting Foundation. (2023). *The Value Reporting Foundation has now consolidated into the IFRS Foundation*. https://www.valuereportingfoundation.org
UN. (2015). *Resolution adopted by the General Assembly on 25 September 2015. Transforming our world: the 2030 Agenda for Sustainable Development*. https://documents-dds-ny.un.org/doc/UNDOC/GEN/N15/291/89/PDF/N1529189.pdf?OpenElement
UNGC. (2018). *Integrating the sustainable development goals into corporate reporting: A practical guide*. https://unglobalcompact.org/library/5628
Wade, M. (2020). *Corporate responsibility in the digital era*. https://sloanreview.mit.edu/article/corporate-responsibility-in-the-digital-era/
Weißenberger, B. E., & Marrocco, A. (2022). Corporate Digital Responsi- bility und Ihre Integration in die Unternehmensführung. In S. Roth & H. Corsten (Eds.), *Handbuch Digitalisierung* (pp. 41–58). Vahlen.

Index

M. Bednárová, K. Soratana (eds.), *Environmental, Social, and Governance (ESG) Investment and Reporting*, https://doi.org/10.1007/978-3-031-84235-1

MIX
Papier aus verantwortungsvollen Quellen
Paper from responsible sources
FSC® C105338

FSC
www.fsc.org

If you have any concerns about our products,
you can contact us on
ProductSafety@springernature.com

In case Publisher is established outside the EU,
the EU authorized representative is:
Springer Nature Customer Service Center GmbH
Europaplatz 3, 69115 Heidelberg, Germany

Printed by Libri Plureos GmbH
in Hamburg, Germany